Introduction to Graph Theory

1-1 BASIC CONCEPTS

Graphs are mathematical objects that can be used to model networks, data structures, process scheduling, computations, and a variety of other systems where the relations between the objects in the system play a dominant role. We will consider graphs from several perspectives: as mathematical entities with a rich and extensive theory; as models for many phenomena, particularly those arising in computer systems; and as structures which can be processed by a variety of sophisticated and interesting algorithms. Our objective in this section is to introduce the terminology of graph theory, define some familiar classes of graphs, illustrate their role in modelling, and define when a pair of graphs are the same.

Terminology.   A *graph* $G(V, E)$ consists of a set V of elements called *vertices* and a set E of unordered pairs of members of V called *edges*. We refer to Figure 1-1 for a geometric presentation of a graph G. The vertices of the graph are shown as points, while the edges are shown as lines connecting pairs of points. The cardinality of V, denoted $|V|$, is called the *order* of G, while the cardinality of E, denoted $|E|$, is called the

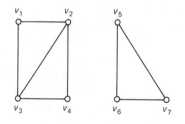

$V(G) = \{v_1, v_2, v_3, v_4, v_5, v_6, v_7\}$

$E(G) = \{(v_1, v_2), (v_2, v_4), (v_2, v_3), (v_4, v_3),$
$\qquad\quad (v_3, v_1), (v_5, v_6), (v_6, v_7), (v_7, v_5)\}$

Order ($| V(G) |$) = 7

Size ($| E(G) |$) = 8

Number of components = 2

Figure 1-1.   Example graph $G(V, E)$.

size of G. When we wish to emphasize the order and size of the graph, we refer to a graph containing p vertices and q edges as a (p, q) graph. Where we wish to emphasize the dependence of the set V of vertices on the graph G, we will write $V(G)$ instead of V, and we use $E(G)$ similarly. The graph consisting of a single vertex is called the *trivial graph*.

We say a vertex u in $V(G)$ is *adjacent to* a vertex v in $V(G)$ if $\{u, v\}$ is an edge in $E(G)$. Following a convention commonly used in graph theory, we will denote the edge between the pair of vertices u and v by (u, v). We call the vertices u and v the *endpoints* of the edge (u, v), and we say the edge (u, v) is *incident with* the vertices u and v. Given a set of vertices S in G, we define the *adjacency set* of S, denoted *ADJ(S)*, as the set of vertices adjacent to some vertex in S. A vertex with no incident edges is said to be *isolated*, while a pair of edges incident with a common vertex are said to be *adjacent*.

The *degree* of a vertex v, denoted by $\deg(v)$, is the number of edges incident with v. If we arrange the vertices of G, $v_1, \ldots, v_n$, so their degrees are in nondecreasing order of magnitude, the sequence $(\deg(v_1), \ldots, \deg(v_n))$ is called the *degree sequence* of G. We denote the *minimum degree* of a vertex in G by *min(G)* and the *maximum degree* of a vertex in G by *max(G)*. There is a simple relationship between the degrees of a graph and the number of edges.

Theorem (Degree Sum). The sum of the degrees of a graph $G(V, E)$ satisfies

$$\sum_{i=1}^{|V|} \deg(v_i) = 2|E|.$$

The proof follows immediately from the observation that every edge is incident with exactly two vertices. Though simple, this result is frequently useful in extending local estimates of the cost of an algorithm in terms of vertex degrees to global estimates of algorithm performance in terms of the number of edges in a graph.

A *subgraph* S of a graph $G(V, E)$ is a graph $S(V', E')$ such that V' is contained in V, E' is contained in E, and the endpoints of any edge in E' are also in V'. A subgraph is said to *span* its set of vertices. We call S a *spanning subgraph* of G if V' equals V. We call S an *induced subgraph* of G if whenever u and v are in V' and (u, v) is in E, (u, v) is also in E'. We use the notation $G - v$, where v is in $V(G)$, to denote the induced subgraph of G on V-$\{v\}$. Similarly, if V' is a subset of $V(G)$, then $G - V'$ denotes the induced subgraph on $V - V'$. We use the notation $G - (u, v)$, where (u, v) is in $E(G)$, to denote the subgraph $G(V, E - \{(u, v)\})$. If we add a new edge (u, v), where u and v are both in $V(G)$ to $G(V, E)$, we obtain the graph $G(V, E \cup \{(u, v)\})$, which we will denote by $G(V, E) \cup (u, v)$. In general, given a pair of graphs $G(V_1, E_1)$ and $G(V_2, E_2)$, their *union* $G(V_1, E_1) \cup G(V_2, E_2)$ is the graph $G(V_1 \cup V_2, E_1 \cup E_2)$. If $V_1 = V_2$ and E_1 and E_2 are disjoint, the union is called the *edge sum* of $G(V_1, E_1)$ and $G(V_2, E_2)$. The *complement* of a graph $G(V, E)$, denoted by G^c, has the same set of vertices as G, but a pair of vertices are adjacent in the complement if and only if the vertices are not adjacent in G.

We define a *path* from a vertex u in G to a vertex v in G as an alternating sequence of vertices and edges,

$$v_1, e_1, v_2, e_2, \ldots, e_{k-1}, v_k,$$

where $v_1 = u$, $v_k = v$, all the vertices and edges in the sequence are distinct, and successive vertices v_i and v_{i+1} are endpoints of the intermediate edge e_i. If we relax the definition to allow repeating vertices in the sequence, we call the resulting structure a *trail*. If we relax the definition further to allow both repeating edges and vertices, we call the resulting structure a *walk*. If we relax the definition of a path to allow the first and last vertices (only) to coincide, we call the resulting closed path a *cycle*. A graph consisting of a cycle on n vertices is denoted by $C(n)$. If the first and last vertices of a trail coincide, we call the resulting closed trail a *circuit*. The *length* of a path, trail, walk, cycle, or circuit is its number of edges.

We say a pair of vertices u and v in a graph are *connected* if and only if there is a path from u to v. We say a graph G is *connected* if and only if every pair of vertices in G are connected. We call a connected induced subgraph of G of maximal order a *component* of G. Thus, a connected graph consists of a single component. The graph in Figure 1-1 has two components. A graph that is not connected is said to be *disconnected*. If all the vertices of a graph are isolated, we say the graph is *totally disconnected*.

A vertex whose removal increases the number of components in a graph is called a *cut-vertex*. An edge whose removal does the same thing is called a *bridge*. A graph with no cut-vertex is said to be *nonseparable* (or *biconnected*). A maximal nonseparable subgraph of a graph is called a *block* (*biconnected component* or *bicomponent*). In general, the *vertex connectivity* (*edge connectivity*) of a graph is the minimum number of vertices (edges) whose removal results in a disconnected or trivial graph. We call a graph G *k-connected* or *k-vertex connected* (*k-edge connected*) if the vertex (edge) connectivity of G is at least k.

If G is connected, the path of least length from a vertex u to a vertex v in G is called the *shortest path* from u to v, and its length is called the *distance* from u to v. The *eccentricity* of a vertex v is defined as the distance from v to the most distant vertex from v. A vertex of minimum eccentricity is called a *center*. The eccentricity of a center of G is called the *radius* of G, and the maximum eccentricity among all the vertices of G is called the *diameter* of G. We can define the *nth power* of a connected graph $G(V, E)$, denoted by G^n as follows: $V(G^n) = V(G)$, and an edge (u, v) is in $E(G^n)$ if and only if the distance from u to v in G is at most n. G^2 and G^3 are called the *square* and *cube* of G, respectively.

If we impose directions on the edges of a graph, interpreting the edges as ordered rather than unordered pairs of vertices, we call the corresponding structure a *directed graph* or *digraph*. In contrast and for emphasis, we will sometimes refer to a graph as an *undirected graph*. We will follow the usual convention in graph theory of denoting an edge from a vertex u to a vertex v by (u, v), leaving it to the context to determine whether the pair is to be considered ordered (directed) or not (undirected). The first vertex u is called the *initial vertex* or *initial endpoint* of the edge, and the second vertex is called the *terminal vertex* or *terminal endpoint* of the edge. If G is a digraph and (u, v) an edge of G, we say the initial vertex u is *adjacent to* v, and the terminal vertex v is *adjacent from* u. We call the number of vertices adjacent to v the *in-degree* of v, denoted indeg(v), and the number of vertices adjacent from v the *out-degree* of v, denoted outdeg(v). The Degree Sum Theorem for graphs has the obvious digraph analog.

Theorem (Digraph Degree Sum). Let $G(V, E)$ be a digraph; then

$$\sum_{i=1}^{|V|} \text{indeg}(v_i) = \sum_{i=1}^{|V|} \text{outdeg}(v_i) = |E|\,.$$

Generally, the terms we have defined for undirected graphs have straightforward analogs for directed graphs. For example, the paths, trails, walks, cycles, and circuits of undirected graphs are defined similarly for directed graphs, with the obvious refinement that pairs of successive vertices of the defining sequences must determine edges of the digraph. That is, if the defining sequence of the directed walk (path, etc.) includes a subsequence v_i, e_i, v_{i+1}, then e_i must equal the directed edge (v_i, v_{i+1}). If we relax this restriction and allow e_i to equal either (v_i, v_{i+1}) or (v_{i+1}, v_i), we obtain a *semiwalk* (*semipath*, *semicycle*, and so on).

A digraph $G(V, E)$ is called *strongly connected*, if there is a (directed) path between every pair of vertices in G. A vertex u is said to be *reachable from* a vertex v in G, if there is a directed path from v to u in G. The digraph obtained from G by adding the edge (v, u) between any pair of vertices v and u in G whenever u is reachable from v (and (v, u) is not already in $E(G)$) is called the *transitive closure* of G.

There are several other common generalizations of graphs. For example, in a *multigraph*, we allow more than one edge between a pair of vertices. In contrast, an ordinary graph that does not allow parallel edges is sometimes called a *simple graph*. In a *loop graph*, both endpoints of an edge may be the same, in which case such an edge is called a *loop* (or *self-loop*). If we allow both undirected and directed edges, we obtain a so-called *mixed graph*.

Special graphs.   Various special graphs occur repeatedly in practice. We will introduce the definitions of some of these here, and examine them in more detail in later sections.

A graph containing no cycles is called an *acyclic graph*. A directed graph containing no directed cycles is called an *acyclic digraph*, or sometimes a Directed Acyclic Graph (DAG). Perhaps the most commonly encountered special undirected graph is the *tree*, which we define as a connected, acyclic graph. An arbitrary acyclic graph is called a *forest*. Thus, the components of a forest are trees. We will consider trees and acyclic digraphs in Chapter 4.

A graph of order N in which every vertex is adjacent to every other vertex is called a *complete graph*, and is denoted by $K(N)$. Every vertex in a complete graph has the same full degree. More generally, a graph in which every vertex has the same, not necessarily full, degree is called a *regular graph*. If the common degree is r, the graph is called *regular of degree r*.

A graph that contains a cycle which spans its vertices is called a *hamiltonian graph*. These graphs are the subject of an extensive literature revolving around their theoretical properties and the algorithmic difficulties involved in efficiently recognizing them. A graph that contains a circuit which spans its edges is called an *eulerian graph*. Unlike hamiltonian graphs, eulerian graphs are easy to recognize. They are merely the connected graphs all of whose degrees are even. We will consider them later in this chapter.

A graph $G(V, E)$ (or $G(V_1, V_2, E)$) is called a *bipartite graph* or *bigraph* if its vertex set V is the disjoint union of sets V_1 and V_2, and every edge in E has the form (v_1, v_2), where $v_1 \in V_1$ and $v_2 \in V_2$. A *complete bipartite graph* is a bigraph in which every vertex in V_1 is adjacent to every vertex in V_2. A complete bigraph depends only on the cardinalities M and N of V_1 and V_2 respectively, and so is denoted by $K(M, N)$. Generally, we say a graph $G(V, E)$ or $G(V_1, \ldots, V_k, E)$ is *k-partite* if the vertex set V is the union of k disjoint sets $V_1, \ldots, V_k$, and every edge in E is of the form (v_i, v_j), for

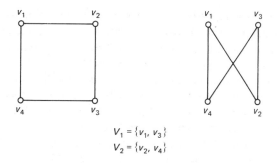

$$V_1 = \{v_1, v_3\}$$
$$V_2 = \{v_2, v_4\}$$

(a) Cyclic and bipartite presentations of $C(4)$.

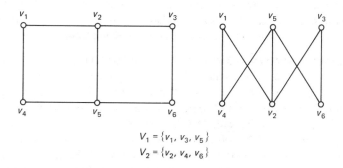

$$V_1 = \{v_1, v_3, v_5\}$$
$$V_2 = \{v_2, v_4, v_6\}$$

(b) Different presentations of a bipartite graph $G(V_1, V_2, E)$.

Figure 1-2. Bipartite and nonbipartite graphs.

vertices $v_i \in V_i$ and $v_j \in V_j$, V_i and V_j distinct. A *complete k-partite graph* is defined similarly. We refer to Figure 1-2 for an example.

Graphs as models. We will describe many applications of graphs in later chapters. The following are illustrative.

Assignment Problem. Bipartite graphs can be used to model problems where there are two kinds of entities, and each entity of one kind is related to a subset of entities of the other. For example, one set may be a set V_1 of employees and the other a set V_2 of tasks the employees can perform. If we assume each employee can perform some subset of the tasks and each task can be performed by some subset of the employees, we can model this situation by a bipartite graph $G(V_1, V_2, E)$, where there is an edge between v_1 in V_1 and v_2 in V_2 if and only if employee v_1 can perform task v_2.

We could then consider such problems as determining the smallest number of employees who can perform all the tasks, which is equivalent in graph-theoretic terms to asking for the smallest number of vertices in V_1 that together are incident to all the vertices in V_2, a so-called covering problem (see Chapter 8). Or, we might want to know how to assign the largest number of tasks, one task per employee and one employee per task, a problem which corresponds graph-theoretically to finding a maximum size regular subgraph of degree one in the model graph, the so-called *matching problem*. Section 1-3 gives a condition for the existence of matchings spanning V_1, and Section 1-5 and Chapters 8 and 9 give algorithms for finding maximum matchings on graphs.

Data Flow Diagrams. We can use directed bipartite graphs to model the flow of data between the operators in a program, employing a *data flow diagram*. For this diagram, we let the data (program variables) correspond to the vertices of one part V_1 of the bigraph $G(V_1, V_2)$, while the operators correspond to the other part V_2. We include an edge from a datum to an operator vertex if the corresponding datum is an input to the corresponding operator. Conversely, we include an edge from an operator vertex to a datum, if the datum is an output of the operator. A bipartite representation of the data flow of a program is shown in Figure 1-3.

A data flow diagram helps in analyzing the intrinsic parallelism of a program. For example, the length of the longest path in a data flow diagram determines the shortest completion time of the program, provided the maximum amount of parallelism is used. In the example, the path $X(+:1)P(\text{DIV}:2)Q(-:4)S(\text{DIV}:6)U$ is a longest path; so the program cannot possibly be completed in less than four steps (sequential operations). A minimum length parallel execution schedule is $\{1\}$, $\{2, 3\}$ concurrently, $\{4, 5\}$ concurrently, and $\{6\}$, where the numbers indicate the operators.

Graph isomorphism.

For any mathematical objects, the question of their equality or identity is fundamental. For example, a pair of fractions (which may look different) are the same if their difference is zero. Just like fractions, a pair of graphs may also look different but actually have the same structure. Thus, the graphs in Figure 1-4 are nominally different because the vertices v_1 and v_2 are adjacent in one of the graphs, but not in the other. However, if we ignore the names or labels on the vertices, the graphs are clearly structurally identical. Structurally identical graphs are called Isomorphic, from the Greek words *iso* for "same" and *morph* for "form." Formally, we define a pair of graphs $G_1(V, E)$ and $G_2(V, E)$ to be *isomorphic* if there is a one-to-one correspondence (or mapping) M between $V(G_1)$ and $V(G_2)$ such that u and v

(1) $P = X + Y$
(2) $Q = Y \text{ div } P$
(3) $R = X * P$
(4) $S = R - Q$
(5) $T = R * P$
(6) $U = T \text{ div } S$

(a) Code sequence.

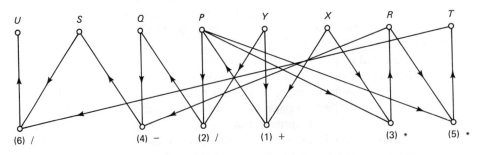

(b) Data flow diagram for (a).

Figure 1-3. Bipartite data flow model.

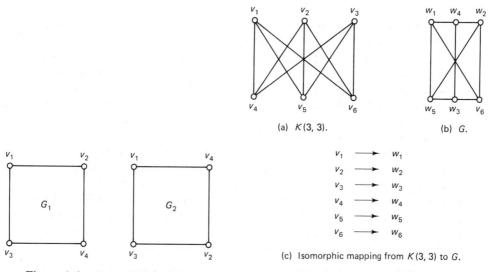

(a) $K(3, 3)$.

(b) G.

$$v_1 \longrightarrow w_1$$
$$v_2 \longrightarrow w_2$$
$$v_3 \longrightarrow w_3$$
$$v_4 \longrightarrow w_4$$
$$v_5 \longrightarrow w_5$$
$$v_6 \longrightarrow w_6$$

(c) Isomorphic mapping from $K(3, 3)$ to G.

Figure 1-4. Isomorphic graphs.

Figure 1-5. A pair of isomorphic graphs.

are adjacent in G_1 if and only if the corresponding vertices $M(u)$ and $M(v)$ are adjacent in G_2. See Figure 1-5 for another example.

To prove a pair of graphs are isomorphic, we need to find an isomorphism between the graphs. The brute force approach is to exhaustively test every possible one-to-one mapping between the vertices of the graphs, until we find a mapping that qualifies as an isomorphism, or determine there is none. See Figure 1-6 for a high-level view of such a search algorithm. Though methodical, this method is computationally infeasible for large graphs. For example, for graphs of order N, there are a priori $N!$ distinct possible 1-1 mappings between the vertices of a pair of graphs; so examining each would be prohibitively costly. Though the most naive search technique can be improved, such as in the backtracking algorithm in Chapter 2, all current methods for the problem are inherently tedious.

A *graphical invariant* (or *graphical property*) is a property associated with a graph $G(V, E)$ that has the same value for any graph isomorphic to G. One may fortui-

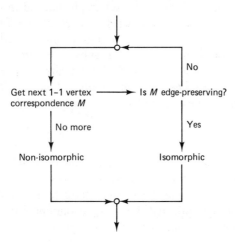

Figure 1-6. Exhaustive search algorithm for isomorphism.

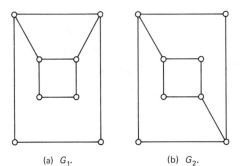

(a) G_1.   (b) G_2.   **Figure 1-7.** G_1 isomorphic to G_2?

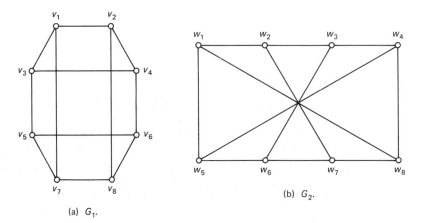

(a) G_1.   (b) G_2.

Figure 1-8. G_1 isomorphic to G_2?

tously succeed in finding an invariant a pair of graphs do not share, thus establishing their nonisomorphism. Of course, two graphs may agree on many invariants and still be nonisomorphic, although the likelihood of this occurring decreases with the number of their shared properties. The graphs in Figure 1-7 agree on several graphical properties: size, order, and degree sequence. However, the subgraph induced in $G_2(V, E)$ by the vertices of degree 2 is regular of degree 0, while the corresponding subgraph in $G_1(V, E)$ is regular of degree 1. This corresponds to a graphical property the graphs do not share; so the graphs are not isomorphic. The graphs in Figure 1-8 also agree on size, order, and degree sequence. However, $G_1(V, E)$ appears to have cycles only of lengths 4, 6, and 8, while $G_2(V, E)$ has cycles of length 5. One can readily show that G_1 is bipartite while G_2 is not, so these graphs are also nonisomorphic.

1-2 REPRESENTATIONS

There are a variety of standard data structure representations for graphs. Each representation facilitates certain kinds of access to the graph but makes complementary kinds of access more difficult. We will describe the simple static representations first. The linked representations are more complicated, but more economical in terms of storage requirements, and they facilitate dynamic changes to the graphs.

1-2-1 Static Representations

The standard static representations are the $|V| \times |V|$ adjacency matrix, the edge list, and the edge incidence matrix.

Adjacency matrix. We define the *adjacency matrix* **A** of a graph $G(V, E)$ as the $|V| \times |V|$ matrix:

$$\mathbf{A}(i,j) = \begin{cases} 1 & \text{if } (i,j) \text{ is in } E(G) \\ 0 & \text{if } (i,j) \text{ is not in } E(G). \end{cases}$$

The adjacency matrix defines the graph completely in $O(|V|^2)$ bits. We can define a corresponding data type Graph_Adjacency or Graph as follows. We assume that the vertices in V are identified by their indices $1..|V|$.

type *Graph* = **record**
|V|: Integer Constant
A(|V|,|V|): 0..1
end

Following good information-hiding practice, we package all the defining information, the order $|V|$ (or $|V(G)|$) included, in one data object.

Let us consider how well this representation of a graph facilitates different graph access operations. The basic graph access operations are

(1) Determine whether a given pair of vertices are adjacent, and

(2) Determine the set of vertices adjacent to a given vertex.

Since the adjacency matrix is a direct access structure, the first operation takes $O(1)$ time. On the other hand, the second operation takes $O(|V|)$ time, since it entails sequentially accessing a whole row (column) of the matrix.

We define the *adjacency matrix representation of a digraph G* in the same way as for an undirected graph:

$$\mathbf{A}(i,j) = \begin{cases} 1 & \text{if } (i,j) \text{ is in } E(G) \\ 0 & \text{if } (i,j) \text{ is not in } E(G). \end{cases}$$

While the adjacency matrix for a graph is symmetric, the adjacency matrix for a digraph is asymmetric.

Matrix operations on the matrix representations of graphs often have graphical interpretations. For example, the powers of the adjacency matrix have a simple graphical meaning for both graphs and digraphs.

Theorem (Adjacency Powers). If **A** is the adjacency matrix of a graph (or digraph) $G(V, E)$, then the (i,j) element of $\mathbf{A}^k$ equals the number of distinct walks from vertex i to vertex j with k edges.

The smallest power k for which $\mathbf{A}^k(i,j)$ is nonzero equals the distance from vertex i to vertex j. Since the expansion of $(\mathbf{A} + \mathbf{I})^k$ contains terms for every positive power of **A** less than or equal to k, its (i,j) entry is nonzero if and only if there is a walk (path) of length k or less from i to j.

The *reachability matrix* $\mathbf{R}$ for a digraph G is defined as

$$\mathbf{R}(i,j) = \begin{cases} 1 & \text{if there is a path from vertex } i \text{ to vertex } j, \\ 0 & \text{otherwise.} \end{cases}$$

We can compute $\mathbf{R}$ in $O(|V|^4)$ operations using the Adjacency Powers Theorem, but can do it much more quickly using methods we shall see later.

Edge list. The *edge list* representation for a graph G is merely the list of the edges in G, each edge being a vertex pair. The pairs are unordered for graphs and ordered for digraphs. A corresponding data type Graph_Edge_List or simply Graph is

> **type** *Graph* = **record**
> |V|: Integer Constant
> |E|: Integer Constant
> Edges(|E|,2): 1..|V|
> **end**

For completeness, we include both the order $|V|$ and the size $|E|$. The array Edges contains the edge list for G. (Edges$(i, 1)$, Edges$(i, 2)$) is the ith edge of G, $i = 1..|E|$.

The storage requirements for the list are $O(|E|)$ bits, or strictly speaking, $O(|E| \log |V|)$ bits since it takes $O(\log |V|)$ bits to uniquely identify one of $|V|$ vertices. If the edges are listed in lexicographic order, vertices i and j can be tested for adjacency in $O(\log |E|)$ steps (assuming comparisons take $O(1)$ time). The set of vertices adjacent to a given vertex v can be found in time $O(\log(|E|) + \deg(v))$, in contrast with the $O(|V|)$ time required for an adjacency matrix.

Incidence matrix. The incidence matrix is a less frequently used representation. Let G be a (p, q) undirected graph. If we denote the vertices by $v_1, \ldots, v_p$ and the edges by $e_1, \ldots, e_q$, we can define the *incidence matrix* $\mathbf{B}$ of G as the $\mathbf{p} \times \mathbf{q}$ matrix:

$$\mathbf{B}(i,j) = \begin{cases} 1 & \text{if vertex } v_i \text{ is incident with edge } e_j, \\ 0 & \text{otherwise.} \end{cases}$$

There is a well-known relationship between the incidence matrix and the adjacency matrix. If we define the *degree matrix* $\mathbf{D}$ of a graph G as the $|V|$ by $|V|$ matrix whose diagonal entries are the degrees of the vertices of G and whose off-diagonal entries are all zero, it is easy to prove the following:

Theorem (Incidence Transpose Product). Let G be a nonempty graph, with incidence matrix $\mathbf{B}$, adjacency matrix $\mathbf{A}$, and degree matrix $\mathbf{D}$. Then $\mathbf{BB}^t = \mathbf{A} + \mathbf{D}$.

We define the *incidence matrix* $\mathbf{B}$ *of the digraph* G as the $\mathbf{p} \times \mathbf{q}$ matrix:

$$\mathbf{B}(i,j) = \begin{cases} +1 & \text{if vertex } v_i \text{ is incident with edge } e_j \text{ and } e_j \text{ is directed out of } v_i, \\ -1 & \text{if vertex } v_i \text{ is incident with edge } e_j, \text{ and } e_j \text{ is directed into } v_i, \\ 0 & \text{otherwise.} \end{cases}$$

The following theorem is easy to establish using the Binet-Cauchy Theorem on the determinant of a product of matrices.

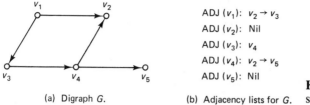

ADJ (v_1): $v_2 \to v_3$	
ADJ (v_2): Nil	
ADJ (v_3): v_4	
ADJ (v_4): $v_2 \to v_5$	
ADJ (v_5): Nil	

(a) Digraph G.   (b) Adjacency lists for G.

Figure 1-9. Adjacency list representation for a digraph.

Theorem (Spanning Tree Enumeration). Let $G(V, E)$ be an undirected connected graph, and let **B** be the incidence matrix of a digraph obtained by assigning arbitrary directions to the edges of G. Then, the number of spanning trees of G equals the determinant of **BB'**.

1-2-2 Dynamic Representations

The *adjacency list* representation for a graph (or digraph) $G(V, E)$ gives for each vertex v in $V(G)$, a list of the vertices adjacent to v. We denote the list of vertices adjacent to v by ADJ(v). Figure 1-9 gives an example. The details of the representation vary depending on how the vertices and edges are represented.

Vertex Representation. The vertices can be organized in

(1) A linear array,

(2) A linear linked list, or

(3) A pure linked structure.

In each case, the set of adjacent vertices ADJ(v) at each vertex v is maintained in a linear linked list; the list header for which is stored in the record representing v. Each of the three types of vertex organization are illustrated for the digraph in Figure 1-9 in Figure 1-10.

In the *linear array* organization, each vertex is represented by a component of a linear array, which acts as the header for the list of edges incident at the vertex. In the *linear list* organization, the vertices are stored in a linked list, each component of which, for a vertex v, also acts as the header for the list of edges in ADJ(v). Refer to Figure 1-10a and b for illustrations.

In the *pure linked* organization, a distinguished vertex serves as an entry vertex to the graph or digraph, and an Entry Pointer points to that entry vertex. The remaining vertices are then accessed solely through the links provided by the edges of the graph. The linked representation for a binary search tree is an example. If not every vertex is reachable from a single entry vertex, a list of enough entry pointers to reach the whole graph is used. Refer to Figure 1-10c for an illustration.

Edge Representation. The representation of edges is affected by whether

(1) The adjacency list represents a graph or digraph, and whether

(2) The representative of an edge is shared by its endpoint vertices or not.

In the case of an undirected graph, the endpoints of an edge (v_i, v_j) play a symmetric role, and so should be equally easily accessible from both ADJ(v_i) and ADJ(v_j). The most natural approach is to represent the edge redundantly under both its end-

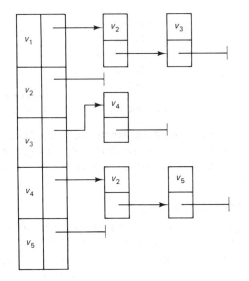

(a) Linear array representation for G.

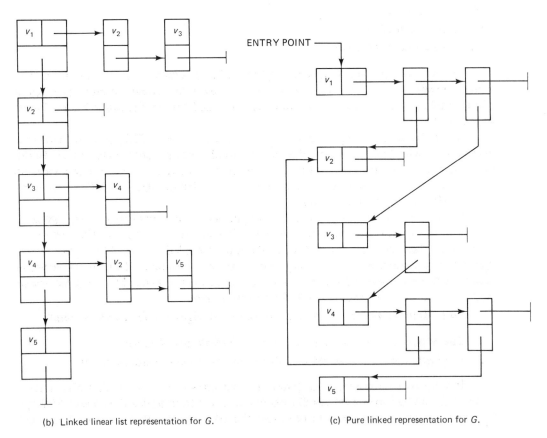

(b) Linked linear list representation for G.

(c) Pure linked representation for G.

Figure 1-10. Different vertex organizations for adjacency lists.

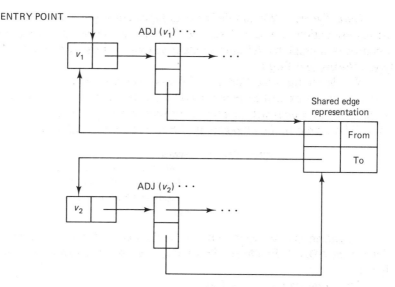

Figure 1-11. Shared representation for pure linked organization.

points, as one does for an adjacency matrix. Alternatively, we can let both lists ADJ(v_i) and ADJ(v_j) share the representative of the edge. The adjacency list entry for each edge then merely points to a separate record which represents the edge. This shared record contains pointers to both its endpoints and any data or status fields appropriate to the edge. Figure 1-11 illustrates this approach.

The adjacency list for a digraph, on the other hand, typically maintains an edge representative (v_i, v_j) on ADJ(v_i) but not on ADJ(v_j). But, if we wish to facilitate quick access to both adjacency-to vertices and adjacency-from vertices, we can use a shared edge representation that lets us access an edge from both ADJ(v_i) and ADJ(v_j). Refer also to Figure 1-11.

Positional Pointers. Many of the graph processing algorithms we will consider process the adjacency lists of the graph in a nested manner. That is, we first process a list ADJ(v_i) partially, up to some edge e_1. Then, we start processing another edge list ADJ(v_j). We process that list up to an edge e_2. But, eventually, we return to processing the list ADJ(v_i), continuing with the next edge after the last edge processed there, e_1. In order to handle this kind of processing, the algorithm must remember for each list the last edge on the list that it processes.

We will define a *positional pointer* at each vertex, stored in the header record for the vertex, to support this kind of nested processing. The positional pointer will initially point to the first edge on its list and be advanced as the algorithm processes the list. Using the positional pointers, we can define a Boolean function Next(G,u,v) which returns in v the next edge or neighbor of u on ADJ(u), and fails if there is none. Successive calls to Next(G,u,v) return successive ADJ(u) elements, starting with the element indicated by the initial position of the positional pointer. Eventually, Next returns the last element. The subsequent call of Next(G,u,v) fails, returning a nil pointer in v and setting Next to False. If we call Next(G,u,v) again, it responds in a cyclic manner and returns a pointer to the head of ADJ(u).

Data Types. We can define data types corresponding to each of the adjacency list representations we have described. We will illustrate this for the linear array representation of a graph by defining a data type **Graph**. First, we will define the constituent types, **Vertex** and **Edge**.

We define the type Vertex as follows. We assume the vertices are indexed from $1..|V|$. Each vertex will be represented by a component of an array $Head(|V|)$, the array components of which will be records of type **Vertex**. The representative of the ith vertex will be stored in the ith position of Head. The type definition is

> **type** *Vertex* = **record**
> Vertex data fields: Unspecified
> Successor: Edge pointer
> Positional Pointer: Edge pointer
> **end**

$Successor(v)$, for each vertex v, is the header for $ADJ(v)$ and points to the first element on $ADJ(v)$. **Positional Pointer**(v) is used for nested processing as we have indicated.

The **Edge** type is defined as

> **type** *Edge* = **record**
> Edge data fields: Unspecified
> Neighboring Vertex: $1..|V|$
> Successor: Edge pointer
> **end**

Neighboring Vertex gives the index of the other endpoint of the edge. Successor just points to the next edge entry on the list.

The overall type **Graph** is then defined as

> **type** *Graph* = **record**
> $|V|$: Integer Constant
> Head($|V|$): Vertex
> **end**

We can define similar data types for the other graph representations.

1-3 BIPARTITE GRAPHS

We described a model for an employee-task assignment problem in Section 1-1 and considered the problem of assigning every employee to a task, with one task per employee and one employee per task. We can analyze this problem using the concept of matching. A *matching M* on a graph $G(V, E)$ is a set of edges of $E(G)$ no two of which are adjacent. A matching determines a regular subgraph of degree one. We say a matching *spans* a set of vertices X in G if every vertex in X is incident with an edge of the matching. We call a matching that spans $V(G)$ a *complete* (*spanning* or *perfect*) *matching*. A 1-1 mapping between employees and tasks in the assignment problem corresponds to a spanning matching.

A classical algorithm for constructing maximum matchings uses *alternating paths,* which are defined as paths whose edges alternate between matching and non-matching edges. Refer to Figure 1-12 for an example. We can use alternating paths in a

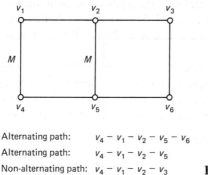

Alternating path:	$v_4 - v_1 - v_2 - v_5 - v_6$	
Alternating path:	$v_4 - v_1 - v_2 - v_5$	
Non-alternating path:	$v_4 - v_1 - v_2 - v_3$	**Figure 1-12.** Alternating paths.

proof of a simple condition guaranteeing the existence of a spanning matching in a bigraph, and hence a solution to the assignment problem.

Theorem (Noncontracting Condition for Existence of Spanning Matchings in Bipartite Graphs). If $G(V_1, V_2, E)$ is bipartite, there exists a matching spanning V_1 if and only if $|\text{ADJ}(S)| \geq |S|$ for every subset S of V_1.

The proof of the theorem is as follows. First, we observe that if there exists a matching spanning V_1, $|\text{ADJ}(S)|$ must be greater than or equal to $|S|$ for every subset S of V_1. We will establish the converse by contradiction.

Suppose the condition is satisfied, but there is no matching that spans V_1. Let M be a maximum matching on G. By supposition, there must be some vertex v in V_1 not incident with M. Define S as the set of vertices in G reachable from v by alternating paths with respect to M; v is the only unmatched vertex in S. Otherwise, we could obtain a matching larger than M merely by reversing the roles of the edges on the alternating path P from v to another unmatched vertex in S. That is, we could change the matching edges on P into nonmatching edges, and vice versa, increasing the size of the matching.

Define W_i ($i = 1, 2$) as the intersection of S with V_i. Since W_1 and W_2 both lie in S and all of S except v is matched, $W_1 - \{v\}$ must be matched by M to W_2. Therefore, $|W_2| = |W_1| - 1$. W_2 is trivially contained in $\text{ADJ}(W_1)$. Furthermore, $\text{ADJ}(W_1)$ is contained in W_2, by the definition of S. Therefore, $\text{ADJ}(W_1)$ must equal W_2, so that $|\text{ADJ}(W_1)| = |W_2| = |W_1| - 1 < |W_1|$, contrary to supposition. This completes the proof.

We next consider the question of how to recognize whether or not a graph is bipartite. The following theorem gives a simple mathematical characterization of bipartite graphs.

Theorem (Bipartite Characterization). A (nontrivial) graph $G(V, E)$ is bipartite if and and only if it contains no cycles of odd length.

The proof of this theorem is straightforward.

Despite its mathematical appeal, the theorem does not provide an efficient way to test whether a graph is bipartite. Indeed, a search algorithm based on the theorem that tested whether every cycle was even would be remarkably inefficient, not because of the difficulty of generating cycles and testing the number of their edges, but because of the sheer volume of cycles that have to be tested. Figure 1-13 overviews an exhaustive

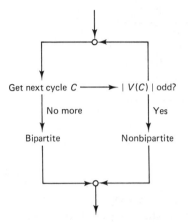

Figure 1-13. Exhaustive search algorithm for testing bipartiteness.

search algorithm that systematically generates and tests every combinatorial object in the graph (here, cycles) which must satisfy a certain property (here, not having odd length) in order for a given property (here, bipartiteness) to be true for the graph as a whole. The performance of the algorithm is determined by how often its search loop is executed, which in turn is determined by how many test objects the graph contains. In this case, the number of test objects (cycles) is typically exponential in the number of vertices in the graph.

Another search algorithm to test for bipartiteness follows immediately from the definition: merely examine every possible partition of $V(G)$ into disjoint parts V_1 and V_2, and test whether or not the partition is a bipartite one. Once again, the difficulty that arises is not in generating and testing the partitions but in examining the exponential number of partitions that occur.

We now describe an efficient bipartite recognition algorithm. The idea of the algorithm is quite simple. Starting with an initial vertex u lying in part V_1, we fan out from u, assigning subsequent vertices to parts (V_1 or V_2) which are determined by the initial assignment of u to V_1. Obviously, the graph is nonbipartite only if some vertex is forced into both parts.

A suitable representation for G is a standard linear array, where the **Header** array is $H(|V|)$ and the array components are records of type

> **type** Vertex = **record**
> Part: 0..2
> Positional Pointer: Edge pointer
> Successor: Edge pointer
> **end**

The entries on the adjacency lists have the form

> **type** Edge = **record**
> Neighbor: 1..|V|
> Successor: Edge pointer
> **end**

We call the type of the overall structure *Graph*.

The function Bipartite uses a utility Get(u) which returns a vertex u for which Part(u) is zero or fails. The outermost loop of the procedure processes the successive components of the graph. The middle loop processes the vertices within a component. The innermost loop processes the immediate neighbors of a given vertex. The algorithm takes time $O(|V| + |E|)$. This is the best possible performance since any possible algorithm must inspect every edge because the inclusion of even a single additional edge could alter the status of the graph from bipartite to nonbipartite.

```
Function Bipartite (G)

(* Returns the bipartite status of G in Bipartite *)

var G: Graph
    Component: Set of 1..|V|
    Bipartite: Boolean function
    v,u: 1..|V|

Set Bipartite to True
Set Part(v)   to 0, for every v ∈ V(G)

while Get(u) do

    Set Component to {u}
    Set Part(u)        to 1

    while Component <> Empty  do

        Remove a vertex v from Component

        for every neighbor w of v  do

            if    Part(w) = 0

            then Add w to Component
                 Set Part(w) to 3 − Part(v)

            else if    Part(v) = Part (w)
                 then Set Bipartite to False
                 Return

End_Function_Bipartite
```

1-4 REGULAR GRAPHS

A graph $G(V, E)$ is *regular of degree r* if every vertex in G has degree r. Regular graphs are encountered frequently in modeling. For example, many parallel computers have interconnection networks which are regular. The five regular polyhedra (the tetrahedron, cube, octahedron, dodecahedron, and icosahedron) also determine regular graphs. The following theorems summarize their basic properties.

Theorem (Regular Graphs).   If $G(V, E)$ is a (p, q) graph which is regular of degree r, then

$$pr = 2q.$$

If G is also bipartite, then

$$|V_1| = |V_2|.$$

Interestingly, every graph G can be represented as an induced subgraph of a larger regular graph.

Theorem (Existence of Regular Supergraph).   Let $G(V, E)$ be a graph of maximum degree M. Then, there exists a graph H which is regular of degree M and that contains G as an induced subgraph.

The following result of Koenig is classical. It is an example of a so-called *factorization* result (see also Chapter 8).

Theorem (Partitioning Regular Bigraphs into Matchings).   A bipartite graph $G(V_1, V_2, E)$ regular of degree r can be represented as an edge disjoint union of r complete matchings.

We can prove this result using the condition for the existence of spanning matchings in bipartite graphs given in Section 1-3. The proof is by induction on the degree of regularity r. The theorem is trivial for r equal to 1. We will assume the theorem is true for degree $r - 1$ or less, and then establish it for degree r.

First, we shall prove that V_1 satisfies the condition of the Bipartite Matching Theorem. Let S be a nonempty subset of V_1 and let $\mathrm{ADJ}(S)$ denote the adjacency set of S. We will show that $|\mathrm{ADJ}(S)| \geq |S|$. Observe that there are $r|S|$ edges emanating from S. By the regularity of G, the number of edges linking $\mathrm{ADJ}(S)$ and S cannot be greater than $r|\mathrm{ADJ}(S)|$. Therefore, $r|\mathrm{ADJ}(S)| \geq r|S|$, or $|\mathrm{ADJ}(S)| \geq |S|$, as required. Therefore, we can match V_1 with a subset of V_2.

Since G is regular, $|V_1| = |V_2|$. Therefore, the matching is actually a perfect matching. We can remove this matching to obtain a reduced regular graph of degree $r - 1$. It follows by induction that $G(V_1, V_2, E)$ can be covered by a union of r edge disjoint matchings. This completes the proof.

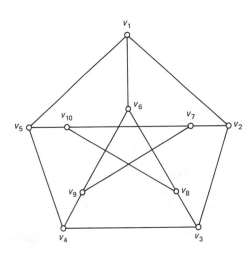

Figure 1-14.   Petersen's graph: The unique five-cage.

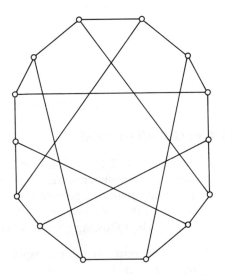

Figure 1-15.   Heawood's graph: The unique six-cage.

Regular graphs are prominent in *extremal graph theory,* the study of graphs that attain an extreme value of some numerical graphical property under a constraint on some other graphical property. An extensively studied class of extremal regular graphs are the *cages.* Let us define the *girth* of a graph G as the length of a shortest cycle in G. Then, an *(r, n)-cage* is an *r*-regular graph of girth n of least order. If *r* equals 3, we call the *(r, n)*-cage an *n-cage.* Alternatively, an *n*-cage is a 3-regular (or *cubic*) graph of girth *n*. Cages are highly symmetric, as illustrated by the celebrated examples of *n*-cages shown in Figure 1-14, the *Petersen graph,* the unique five-cage, and in Figure 1-15, the *Heawood graph,* the unique six-cage.

1-5 MAXIMUM MATCHING ALGORITHMS

There are a number of interesting algorithms for finding maximum matchings on graphs. This section describes four of them, each illustrating a different methodology: maximum network flows, alternating paths, integer programming, or randomization.

1-5-1 Maximum Flow Method (Bigraphs)

A maximum matching on a bipartite graph can be found by modeling the problem as a network flow problem and finding a maximum flow on the model network. The theory of network flows is elaborated in Chapter 6. Basically, a flow network is a digraph whose edges are assigned capacities, and each edge is interpreted as a transmission link capable of bearing a limited amount of "traffic flow" in the direction of the edge. One can then ask for the maximum amount of traffic flow that can be routed between a pair of vertices in such a network. This maximum traffic flow value and a set of routes that realize it can be found using maximum flow algorithms. Efficient maximum flow algorithms for arbitrary graphs are given in Chapter 6.

To model the bipartite matching problem as a maximum network flow problem, we can use the kind of flow network shown in Figures 1-16 and 1-17. We transform the

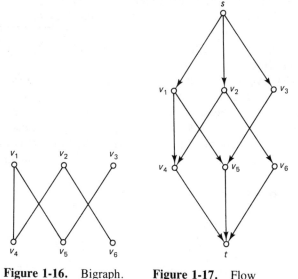

Figure 1-16. Bigraph.   **Figure 1-17.** Flow network with unit capacity edges.

bigraph to a flow network by introducing vertices s and t, which act as the source and sink of the traffic to be routed through the network; we then direct the edges of the bigraph downwards and assign a capacity limit of one unit of traffic flow per edge. A flow maximization algorithm is then used to find the maximum amount of traffic that can be routed between s and t. The internal (bipartite) edges of the flow paths determine a maximum matching on the original bigraph. In the example, the traffic is routed on the paths s-v_1-v_4-t, s-v_2-v_6-t, and s-v_3-v_5-t, one unit of traffic flow per path. The corresponding maximum matching is (v_1, v_4), (v_2, v_6), and (v_3, v_5).

1-5-2 Alternating Path Method (Bigraphs)

The method of alternating paths is a celebrated search tree technique for finding maximum matchings. A general method for arbitrary graphs is given in Chapter 8, but it simplifies greatly when the graph is bipartite, which is the case considered here.

Let $G(V, E)$ be a graph and let M be a matching on G. A path whose successive edges are alternately in M and outside of M is called an *alternating path with respect to M*. We call a vertex not incident with an edge of M a *free* (or *exposed*) *vertex relative to M*. An alternating path with respect to M whose first and last vertices are free is called an *augmenting path with respect to M*. We can use these paths to transform an existing matching into a larger matching by simply interchanging the matching and nonmatching edges on an augmenting path. The resulting matching has one more edge than the original matching. Refer to Figure 1-18 for an example.

Augmenting paths can be used to enlarge a matching; if they do not exist, a matching must already be maximum.

Theorem (Augmenting Path Characterization of Maximum Matchings). A matching M on a graph $G(V, E)$ is a maximum matching if and only if there is no augmenting path in G with respect to M.

The proof follows. We show that if M is not a maximum matching there must exist an augmenting path in G between a pair of free vertices relative to M. The proof is by contradiction. Suppose that M' is a larger matching than M. Let G' denote the symmetric difference of the subgraphs induced by M and M'. That is, the edges of G' are the edges of G which are in either M or M' but not both. By supposition, there are more edges in G' from M' than from M.

Observe that every vertex in G' has degree at most two (in G') since, at most, two edges, one from M and one from M', are adjacent to any vertex of G'. Therefore, the components of G' are either isolated vertices, paths, or cycles. Furthermore, since none of the edges of a given matching are mutually adjacent, each path or cycle component of G' must be an alternating path. In particular, any path of even length and every cycle (all of which are necessarily of even length) must contain an equal number of

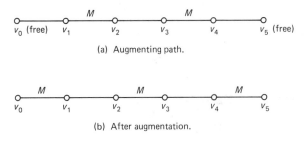

(a) Augmenting path.

(b) After augmentation.

Figure 1-18. Application of an augmenting path.

edges from M and M'. Therefore, if we remove every cycle and every component consisting of an even length path from G', there still remain more edges of M' than of M. Therefore, at least one of the remaining path components must have more edges from M' than M, indeed exactly one more edge. Since the edges of this path alternate between M and M', its first and last edges must lie in M'. Consequently, neither the first nor last vertex of the path can lie in M. Otherwise, the path could be extended, contrary to the maximality of a component. Thus, the component must be an alternating path with respect to M with free endpoints, that is, an augmenting path with respect to M. This completes the proof.

The Characterization Theorem suggests a procedure for finding a maximum matching. We start with an empty matching, find an augmenting path with respect to the matching, invert the role of the edges on the path (matching to nonmatching, and vice versa), and repeat the process until there are no more augmenting paths, at which point the matching is maximum. The difficulty lies in finding the augmenting paths. We shall show how to find them using a search tree technique. The technique is extremely complicated when applied to arbitrary graphs, but simplifies in the special case of bipartite graphs.

Alternating search tree. The augmenting path search tree consists of alternating paths emanating from a free vertex (Root), like the tree shown in Figure 1-19. The idea is simply to repeatedly extend the tree until it reaches a free vertex v (other than Root). If the search is successful, the alternating path through the tree from Root to v is an augmenting path. Thus, in the example the tree path from Root to v_9 is augmenting. We can show, conversely, that if the search tree does not reach a free vertex, there is no augmenting path (with Root as a free endpoint) in the graph. The blocked

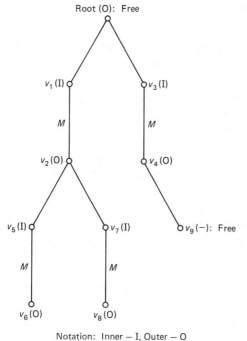

Notation: Inner — I, Outer — O

Figure 1-19. Search tree.

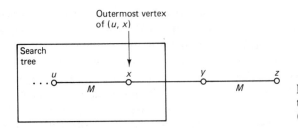

Figure 1-20. Extension of search tree using the pair of edges (x, y) $\cup$ (y, z).

search tree is called a *Hungarian tree,* and its vertices can be ignored in all subsequent searches. When the search trees at every free vertex become blocked, there are no augmenting paths at all, and so the matching is maximum.

We construct the search tree by extending the tree two alternating edges at a time, maintaining the alternating character of the paths through the search tree (as required in order to eventually find an augmenting alternating path) by always making the extensions either from the root of the tree or from the outermost vertex of some matching edge. That is, suppose that (u, x) is a matching edge already lying in the search tree and that x is the endpoint of (u, x) lying farthest from the root of the search tree, as illustrated in Figure 1-20. Denote an unexplored edge at x by (x, y). If y is free, the path through the tree from Root to y is an augmenting path, and the search may terminate. Otherwise, if y is a matched vertex, we let its matching edge be (y, z). Then, we can extend the search tree by adding the pair of alternating edges (x, y) and (y, z) to the tree.

We will refer to search tree vertices as either *outer* or *inner vertices* according to whether their distance, through the search tree, from the root is even or odd, respectively. The tree advances from outer vertices only. If we construct a search tree following these guidelines, the tree will find an augmenting path starting at the exposed vertex Root if there is one, though not necessarily a given augmenting path. We state this as a theorem.

Theorem (Search Tree Correctness). Let $G(V, E)$ be an undirected graph and let Root be an exposed vertex in G. Then, the search tree rooted at Root finds an augmenting path at Root if one exists.

To prove this, we argue as follows. Suppose there exists an augmenting path P from Root to an unmatched vertex v, but that the search tree becomes blocked before finding any augmenting path from Root, including P. We will show this leads to a contradiction.

Let z be the last vertex on P, starting with v as the first, which is not in the search tree; z exists since by supposition v is not in the search tree. Let w be the next vertex after z on P, proceeding in the direction from v to Root; w must lie in the search tree, and so must all the vertices on P between w and Root, though not necessarily all the edges of P between w and Root. Furthermore, (w, z) cannot be a matching edge, since every tree vertex is already matched except for Root, which is free. Therefore, (w, z) is an unmatched edge. We will distinguish two cases according to whether w is an outer or an inner vertex.

1. In the case that w is an outer vertex, the search procedure would eventually extend the search tree from w along the unmatched edge (w, z) to z, unless another augmenting path had been detected previously, contradicting the assumption that z is not in the search tree.

2. We will show the case where w is an inner vertex cannot occur. For, since P is alternating, the tree matching edge at w, that is (w, w_1), would be the next edge of P. Therefore, P would necessarily continue through the tree in the direction of (w, w_1). Subsequently, whenever P entered an inner vertex, necessarily via a nonmatching edge, P would have to exit that vertex by the matching tree edge at the vertex, since P is alternating. Furthermore, if P exits the tree at any subsequent point, it must do so for at most a single edge, only from an outer vertex and only via an unmatched edge. If we denote such an edge by (x, y) and x is the vertex where P exits and y is the vertex where P reenters the tree, y cannot also be an outer vertex. For otherwise, the path through the search tree from z, the common ancestor of x and y, to x, plus the edge (x, y), plus the path through the search tree from y to z would together constitute an odd cycle. But, since the graph is a bigraph, it contains no odd cycles by our Characterization Theorem for bigraphs. Consequently, the reentry vertex y must be an inner vertex. Therefore, just as before, P must leave y along the matching edge in the tree which is incident at y. This pattern continues indefinitely, preventing P from ever reaching Root, which is contrary to the definition of P as an augmenting path. This completes the proof of the theorem.

The Hungarian Trees produced during the search process have the following important property.

Theorem (Hungarian Trees). Let $G(V, E)$ be an undirected graph. Let H be a Hungarian tree with respect to a matching M on G rooted at a free vertex Root. Then, H may be ignored in all subsequent searches. That is, a maximum matching on G is the union of a maximum matching on H and a maximum matching on $G - H$.

The proof is as follows. Let M_1 be a maximum matching for $G - H$, and suppose that M' is an arbitrary matching for G, equal to $M_1' \cup M_H' \cup M_I$, where M_1' is in $G - H$, M_H' is in H, and M_I satisfies that the intersection of M_I and $(G - H) \cup H$ is empty. By supposition, $|M_1| \geq |M_1'|$. Furthermore, since every edge in M_I is incident with at least one inner vertex of H; then if I' is the set of inner vertices in H which are incident with M_I, then $|M_I|$ is at most $|I'|$. Finally, observe that $H - I'$ consists of $|I'| + 1$ disjoint alternating trees whose inner vertices together comprise $I - I'$, where I denotes the set of inner vertices of H. Since the cardinality of a maximum matching in an alternating tree equals the number of its inner vertices, it follows that $|M_H'|$ is at most $|I - I'|$. Combining these results, we obtain that $|M'| \leq |M_1'| + |M_H'| + |M_I| \leq |M_1| + |I - I'| + |I'| = |M_1| + |I| = |M_1 \cup M_H|$, which completes the proof.

Maximum matching algorithm. A procedure Maximum_Matching constructs the matching. Its hierarchical structure is

```
Maximum_Matching (G)
    Next_Free (G,Root)
    Find_Augmenting_Tree (G,Root,v)
        Create (Q)
        Enqueue (w,Q)
        Next (G,Head(Q),v)
    Apply_Augmenting_Path (G,Root,v)
    Remove_Tree (G)
    Clear(G)
```

Next_Free returns a free vertex Root, or fails. The procedure Find_Augmenting_ Tree then tries to construct an augmenting search tree rooted at Root. If Find_ Augmenting_Tree is successful, the free vertex at the terminus of the augmenting path is returned in v, and Apply_Augmenting_Path uses the augmenting path to change the matching by following the search tree (predecessor) pointers from the breakthrough vertex v back to Root, modifying the existing matching accordingly. If Find_ Augmenting_Tree fails, the resulting blocked search tree is by definition a Hungarian tree, and we ignore its vertices during subsequent searches, since they cannot be part of any augmenting path.

We have used a queue based search procedure (called *Breadth-First Search,* see Chapter 5) to control the search order, whence the utilities Create (Q) and Enqueue (w,Q). However, other search techniques such as depth-first search could have been used. Find_Augmenting_Tree uses Next(G,u,v) to return the next neighbor v of u, and fails when there is none. The utility Clear (G) is used to reset Pred and Positional Pointer back to their initial values for every vertex in G. Remove_Tree (G) removes from G the subgraph induced by a current Hungarian tree in G. The other utilities are familiar. Refer to Figure 1-21 for an example, where an initial nontrivial matching is given in Figure 1-21a.

The outer-inner terminology used to describe the search process is not explicit in the algorithm. In Find_Augmenting_Tree, w is an outer vertex. These are the only vertices queued for subsequent consideration for extending the search tree. The vertices named v in Find_Augmenting_Tree correspond to what we have called inner vertices. The tree is not extended from them so they are not queued.

We will now describe the data structures for the algorithm. We represent the graph $G(V, E)$ as a linear list with vertices indexed $1..|V|$. The components of the list are of type Vertex. Both the search tree and the matching are embedded in the repre-

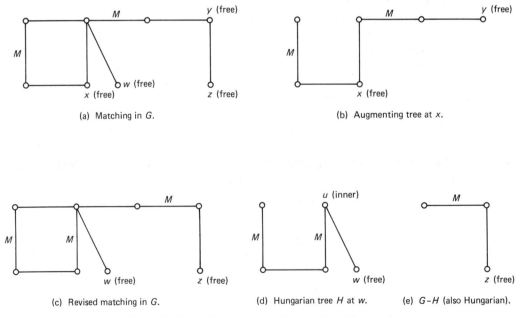

(a) Matching in G.   (b) Augmenting tree at x.

(c) Revised matching in G.   (d) Hungarian tree H at w.   (e) $G-H$ (also Hungarian).

Figure 1-21. Maximum matching on bipartite graph.

sentation of G. Each vertex component heads the adjacency list for the vertex. The type definition is as follows:

```
type   Vertex = record
                  Matching vertex: 0..|V|
                  Pred: Vertex pointer
                  Positional Pointer,
                  Successor: Edge_Entry pointer
              end
```

Matching vertex gives the index of the matching vertex for the given vertex; it is 0 if there is none. Pred points to the predecessor of the given vertex in the search tree or is nil. Pred(v) is nil until v is scanned; so it can be used to detect whether a vertex was scanned previously. Positional Pointer and Successor serve their customary roles as explained in Section 1-2.

We represent the adjacency lists as doubly linked lists with the representative for each edge shared by its endpoints. This organization facilitates the deletion of vertices and edges required when Hungarian trees are deleted from G. The records on the adjacency lists are of type Edge_Entry. Each of these points to its list predecessor, successor, and its associated edge representative. The edge representative itself is of type Edge and is defined by

```
type   Edge = record
                Endpoints(2): 1..|V|
                Endpoint-pointer(2): Edge_Entry pointer
            end
```

Endpoints(2) gives the indices of the endpoints of the represented edge; while Endpoint-pointer(2) points to the edge entries for these endpoints. We call the overall representation for the graph type Graph. Q packages a queue whose entries identify outer vertices to be processed.

The formal statements of Maximum_Matching and Find_Augmenting_Tree are as follows.

```
Procedure Maximum_Matching (G)

(* Finds a maximum matching in G *)

var G: Graph
    Root, v: Vertex pointer
    Next_Free, Find_Augmenting_Tree: Boolean function

while Next_Free (G, Root)  do

    if    Find_Augmenting_Tree (G, Root,v)

    then Apply_Augmenting_Path (G, Root,v)
         Clear (G)

    else  Remove_Tree (G)

End_Procedure_Maximum_Matching
```

Function_Find_Augmenting_Tree (G,Root,v)

(* Tries to find an augmenting tree at Root, returning the
augmenting vertex in v and fails if there is none. *)

var G: Graph
    Root, v, w: Vertex pointer
    Q: Queue
    Empty, Next, Find_Augmenting_Tree: Boolean function
    Dequeue: Vertex pointer function

Set Find_Augmenting_Tree to False
Create (Q); Enqueue (Root, Q)

repeat

  while Next(G,Head(Q),v) **do**

      if v Free **then** (* Augmenting path found *)
               Set Pred(v) to Head(Q)
               Set Find_Augmenting_Tree to True
               return

          else **if** v unscanned

               then (* Extend search tree *)
                    Let w be vertex matched with v
                    Set Pred(w) to v
                    Set Pred(v) to Head(Q)
                    Enqueue(w,Q)

  until Empty(Dequeue(Q))

End_Function_Find_Augmenting_Tree

The performance of the matching algorithm is summarized in the following theorem.

Theorem (Matching Algorithm Performance). Let $G(V, E)$ be an undirected graph. Then, the search tree matching algorithm finds a maximum matching on G in $O(|V||E|)$ steps.

The proof of the theorem is as follows. The performance statement depends strongly on the Hungarian Trees Theorem. Observe that there are at most $|V|/2$ search phases, where a search phase consists of generating successive search trees, each rooted at a different free vertex, until either one search tree succeeds in finding an augmenting path, or every search tree is blocked. Each such search phase may generate many search trees. Nonetheless, since the Hungarian trees can be ignored in all subsequent searches (even during the same phase), at most $|E|$ edges will be explored during a phase, up to the point where one of the searches succeeds or every search fails. Thus, each search phase takes time at most time $O(|V| + |E|)$; so the algorithm takes $O(|V||E|)$ steps overall. This completes the proof.

1-5-3 Integer Programming Model

Integer programming is a much less efficient technique than the previous methods for finding maximum matchings, but it has the advantage of being easy to apply and readily adaptable to more general problems, such as weighted matching, where the objective is to find a maximum weight matching on a weighted graph. The method can also be trivially adapted to maximum degree-constrained matching, where the desired subgraph need not be regular of degree one but only of bounded degree. An application of degree-constrained matching is given in Chapter 8. Despite the simplicity of the method, it suffers from being NP-Complete (see Chapter 10 for this terminology).

The idea is to use a system of linear inequalities for 0-1 variables to model the matching problem, for an arbitrary graph, and then apply the techniques of 0-1 integer programming to solve the model. Let $G(V, E)$ be a graph for which a maximum matching is sought and let A_{ij} be the adjacency matrix of G. Let x_{ij}, i, j $(j > i) = 1, \ldots, |V|$ denote a set of variables which we constrain to take on only 0 or 1 as values. We also constrain the variables; so they behave like indicators of whether an edge is in the matching or not. That is, we restrict each x_{ij} so that it is 0 if (i, j) is not a matching edge and is 1 if (i, j) is a matching edge. Finally, we define an objective function whose value equals the number of edges in the matching and which is maximized subject to the constraints. The formulation is as follows.

Edge Constraints

$$x_{ij} \leq A_{ij}, \quad \text{for } i, j \ (j > i) = 1, \ldots, |V|$$

Matching Constraints

$$\sum_{j>i}^{|V|} x_{ij} + \sum_{j<i}^{|V|} x_{ji} \leq 1, \quad \text{for } i = 1, \ldots, |V|$$

Objective Function (Matching size)

$$\sum_{i=1}^{|V|} \left(\sum_{j>i} x_{ij} \right)$$

The edge constraints ensure only edges in the graph are allowed as matching edges. The matching constraints ensure there is at most one matching edge at each vertex, as required by the graphical meaning of a matching. The objective function just counts the number of edges in the matching.

Figure 1-22 gives a simple example. The system of inequalities for the example is as follows.

Edge Constraints

$$x_{12} \leq 1, \qquad x_{13} \leq 1, \qquad x_{23} \leq 1 .$$

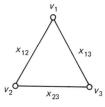

Figure 1-22. Maximum matching problem.

Matching Constraints

$$x_{12} + x_{13} \leq 1, \qquad x_{12} + x_{23} \leq 1, \qquad x_{23} + x_{13} \leq 1.$$

Objective Function

$$x_{12} + x_{13} + x_{23}$$

Three solutions are possible: $x_{12} = 1$, $x_{13} = 0$, $x_{23} = 0$, and the two symmetric variations of this. Each corresponds to a maximum matching.

1-5-4 Probabilistic Method

Surprisingly, combinatorial problems can sometimes be solved faster and more simply by random or probabilistic methods than by deterministic methods. The algorithm we describe here converts the matching problem into an equivalent matrix problem, which is then "solved" by randomizing. The method is due to Lovasz and is based on the following theorem of Tutte.

Theorem (Tutte Matrix Condition For Perfect Matching). Let $G(V, E)$ be an undirected graph. Define a matrix $\mathbf{T}$ of indeterminates as follows: for every pair (i, j) $(j > i) = 1, \ldots, |V|$, for which (i, j) is an edge of G, set $\mathbf{T}(i, j)$ to an indeterminate $+x_{ij}$ and set $\mathbf{T}(j, i)$ to an indeterminate $-x_{ij}$; otherwise, set $\mathbf{T}(i, j)$ to 0. Then, G has a complete matching if and only if the determinant of $\mathbf{T}$ is not identically zero.

The matrix $\mathbf{T}$ is called the *Tutte matrix* of G and its determinant is the *Tutte determinant*. The theorem provides a simple test for the existence of a complete matching in a graph. However, there are computational difficulties involved in applying the theorem.

It is usually straightforward to calculate a determinant. For example, Gaussian Elimination can be used to reduce the matrix to upper triangular form, and the determinant is then just the product of the diagonal entries of the upper triangular matrix. This procedure has complexity $O(|V|^3)$. However, Gaussian Elimination cannot be used when the matrix entries are symbolic, as is the case here. In order to evaluate a symbolic, as opposed to a numerical determinant, we apparently must fall back on the original definition of a determinant. Recall that by definition a determinant of a matrix $\mathbf{T}$ is a sum of $|V|!$ terms.

$$\sum_{\text{all permutations}} \text{sign}(i_1, \ldots, i_{|V|}) \mathbf{T}(1, i_1) * \ldots * \mathbf{T}(|V|, i_{|V|}),$$

where $(i_1, \ldots, i_{|V|})$ is a permutation of the indices $1, \ldots, |V|$ and $\text{sign}(x)$ returns the sign of the permutation x: $+1$ if x is an even permutation and -1 if x is an odd permutation. To apply the Tutte condition entails testing if the symbolic polynomial that results from the evaluation of the determinant is identically zero or not. This apparently simple task is really quite daunting. The polynomial has $|V|!$ terms, which is exponential in $|V|$. Therefore, even though the required evaluation is trivial in a conceptual sense, it is very nontrivial in the computational sense.

Observe that we do not actually need to know the value of the determinant, only whether it is identically zero or not. This suggests the following alternative testing procedure: Perform the determinant condition test by randomly assigning values to the symbolic matrix entries and then evaluating the resulting determinant by a numerical technique like Gaussian Elimination.

If the determinant randomly evaluates to nonzero, then the original symbolic determinant must be nonzero; so the graph must have a complete matching. If the determinant evaluates to zero, there are two possibilities. Either it is really identically zero or it is not identically zero, but we have randomly (accidentally) stumbled on a root of the Tutte determinant. Of course, by repeating the random assignment and numerical evaluation process a few times, we can make the risk of accidentally picking a root negligible. In particular, if the determinant randomly evaluates to zero several times, it is almost certainly identically zero. Thus, we can with great confidence interpret a zero outcome as implying that the graph does not have a complete matching. Thus, positive conclusions (that is, a complete matching exists) are never wrong, while negative conclusions (that is, a complete matching does not exist) are rarely wrong.

The random approach allows us to apply the determinant condition efficiently and with a negligible chance of error. The following procedure Prob_Matching_Test (G) formalizes this approach. Because a subsequent procedure that invokes Prob_Matching_Test will make vertex and edge deletions from the graph, we will use a linear list representation for G instead of the simpler adjacency matrix representation. The vertex type is

```
type   Vertex = record
                Index: 1..|V|
                Adjacency_Header: Edge pointer
                Successor: Vertex pointer
       end
```

G itself is just a pointer to the head of the vertex list. Each vertex has an index field for identification, an Adjacency_Header field, which points to the first element on its edge list, and a Successor field, which points to the succeeding vertex on the vertex list. We can assume the vertex records are ordered according to index. We will use adjacency lists structured in the same manner as for the alternating paths algorithm to facilitate deletions that will be required in a later algorithm.

```
Function Prob_Matching_Test (G)

(* Succeeds if G has a perfect matching, and fails otherwise *)

var G: Vertex pointer
    X(|V|,|V|), T(|V|,|V|): Integer
    R: Integer Constant
    Prob_Matching_Test: Boolean function

Set T to a Tutte matrix for G
Set Prob_Matching_Test to False

repeat   R   Times

    Set X to a random instance of T

    if    Determinant (X) <> 0
    then Set Prob_Matching_Test to True
         return

End_Function_Prob_Matching_Test
```

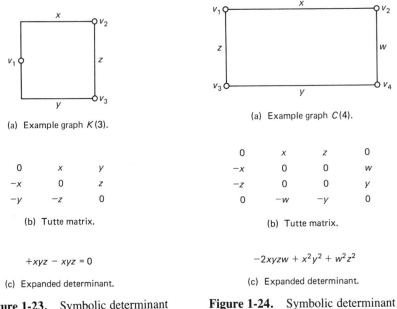

(a) Example graph $K(3)$.

(a) Example graph $C(4)$.

$$\begin{array}{ccc} 0 & x & y \\ -x & 0 & z \\ -y & -z & 0 \end{array}$$

$$\begin{array}{cccc} 0 & x & z & 0 \\ -x & 0 & 0 & w \\ -z & 0 & 0 & y \\ 0 & -w & -y & 0 \end{array}$$

(b) Tutte matrix.

(b) Tutte matrix.

$$+xyz - xyz = 0$$

$$-2xyzw + x^2y^2 + w^2z^2$$

(c) Expanded determinant.

(c) Expanded determinant.

Figure 1-23. Symbolic determinant example 1.

Figure 1-24. Symbolic determinant example 2.

Prob_Matching_Test(G) determines whether or not G has a complete matching, but does not show how to find a complete matching if one exists. But, we can easily design a constructive matching algorithm (for graphs with complete matchings) using this existence test. The following procedure Prob_Complete_Matching returns a complete matching if one exists. We assume that M is initially empty. The representations are the same as before. Figures 1-23 and 1-24 illustrate the technique.

Procedure Prob_Complete_Matching(G, M)

(* Returns a complete matching in M, or the empty set if there
    is none *)

var G: Vertex pointer
      x: Edge
      M: Set of Edge
      Prob_Matching_Test: Boolean function

if   Prob_Matching_Test(G)

then Select an edge x in G
      Set G to G − x

   if   Prob_Matching_Test (G)

   then Prob_Complete_Matching (G, M)

   else **Set** M to M ∪ {x}
         Set G to G − {Endpoints of x}
         Prob_Complete_Matching (G, M)

End_Procedure_Prob_Complete_Matching

1-6 PLANAR GRAPHS

A graph $G(V, E)$ is called a *planar graph* if it can be drawn or embedded in the plane in such a way that the edges of the embedding intersect only at the vertices of G. Figures 1-25 and 1-26 show planar and nonplanar graphs. Both planar and nonplanar embeddings of $K(4)$ are shown in Figure 1-25, while a partial embedding of $K(3, 3)$ is shown in Figure 1-26. One can prove that $K(3, 3)$ is nonplanar; so this embedding cannot be completed. For example, the incompletely drawn edge x in the figure is blocked from completion by existing edges regardless of whether we try to draw it through region I or region II. Of course, this only shows that the attempted embedding fails, not that no embedding of $K(3, 3)$ is possible, though in fact none is. This section describes the basic theoretical and algorithmic results on planarity. We begin with a simple application.

Model: Facility layout. Consider the following architectural design problem. Suppose that n simple planar areas $A_1, \ldots, A_n$ (of flexible size and shape) are to be laid out next to each other but subject to the constraint that each of the areas be adjacent to a specified subset of the other areas. The adjacency constraints can be represented using a graph $G(V, E)$ whose vertices represent the given areas A_i and such that a pair of vertices are adjacent in G if and only if the areas they represent are supposed to be adjacent in the layout. The planar layout can be derived as follows.

(1) Construct a planar representation R of $G(V, E)$.

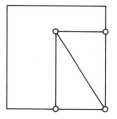

(a) Nonplanar embedding of $K(4)$.　　　(b) Planar embedding of $K(4)$.

Figure 1-25.　Different embeddings of $K(4)$.

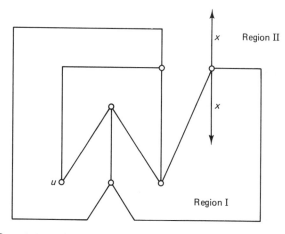

Figure 1-26.　Attempted planar embedding of $K(3, 3)$.

(2) Using R, construct the *planar dual* G' of G as follows:
  (i) Corresponding to each planar region r of R, define a vertex v_r of G', and
  (ii) Let a pair of vertices v_r and v'_r in G' be adjacent if the pair of corresponding regions r and r' share a boundary edge in R.
(3) Construct a planar representation R' of G'.

If step (1) fails, G is nonplanar and the constraints are infeasible. Otherwise, steps (1) and (2) succeed. (Strictly speaking, in step (2ii), if r and r' have multiple boundary edges in common, we connect v_r and v'_r multiply; and if an edge lies solely in one region r, we include a self-loop at v_r in G'. Neither case occurs here.) It can be shown that if G is planar, its dual G' is also; so the planar representation required by step (3) exists. The representation R' is the desired planar layout of the areas $A_1, \ldots, A_n$. The regions of R' correspond to the given areas, and the required area adjacency constraints are satisfied. The process is illustrated in Figure 1-27.

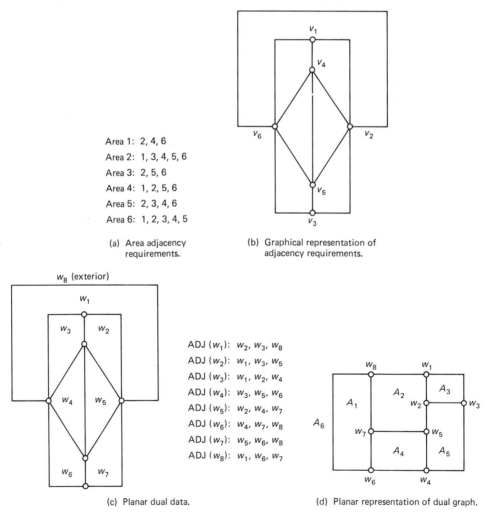

Area 1: 2, 4, 6
Area 2: 1, 3, 4, 5, 6
Area 3: 2, 5, 6
Area 4: 1, 2, 5, 6
Area 5: 2, 3, 4, 6
Area 6: 1, 2, 3, 4, 5

(a) Area adjacency requirements.

(b) Graphical representation of adjacency requirements.

ADJ (w_1): w_2, w_3, w_8
ADJ (w_2): w_1, w_3, w_5
ADJ (w_3): w_1, w_2, w_4
ADJ (w_4): w_3, w_5, w_6
ADJ (w_5): w_2, w_4, w_7
ADJ (w_6): w_4, w_7, w_8
ADJ (w_7): w_5, w_6, w_8
ADJ (w_8): w_1, w_6, w_7

(c) Planar dual data.

(d) Planar representation of dual graph.

Figure 1-27. Planar dual and adjacency constraints.

Planarity testing. There are two classical tests for planarity. The Theorem of Kuratowski characterizes planar graphs in terms of subgraphs they are forbidden to have, while an algorithm of Hopcroft and Tarjan determines planarity in linear time and shows how to draw the graph if it is planar.

Kuratowski's Theorem uses the notion of homeomorphism, a generalization of isomorphism. We first define *series insertion* in a graph $G(V, E)$ as the replacement of an edge (u, v) of G by a pair of edges (u, z) and (z, v), where z is a new vertex of degree two. That is, we insert a new vertex on an existing edge. We define *series deletion* as the replacement of a pair of existing edges (u, z) and (z, v), where z is a current vertex of degree two, by a single edge (u, v), and delete z. That is, we "smooth away" vertices of degree two. Then, a pair of graphs G_1 and G_2 are said to be *homeomorphic* if they are isomorphic or can be made isomorphic through an appropriate sequence of series insertions and/or deletions.

Theorem (Kuratowski's Forbidden Subgraph Characterization of Planarity). A graph $G(V, E)$ is planar if and only if G contains no subgraph homeomorphic to either $K(5)$ or $K(3, 3)$.

We omit the difficult proof of the theorem. Figures 1-28 and 1-29 illustrate its application. The graph in Figure 1-28 is Petersen's graph. Though this graph does not contain a subgraph isomorphic to either $K(5)$ or $K(3, 3)$, it does contain the homeomorph of $K(3, 3)$ shown in Figure 1-29; so it is nonplanar.

Kuratowski's Theorem suggests an exhaustive search algorithm for planarity based on searching for subgraphs homeomorphic to one of the forbidden subgraphs. But, it is not obvious how to do this in polynomial time since the forbidden homeomorphs can have arbitrary orders (see the exercises). Williamson (1984) describes a fairly complicated algorithm based on depth-first search that finds a Kuratowski subgraph in a nonplanar graph in $O(|V|)$ time.

It is sometimes simpler to apply Kuratowski's planarity criterion in a different form. We define *contraction* as the operation of removing an edge (u, v) from a graph

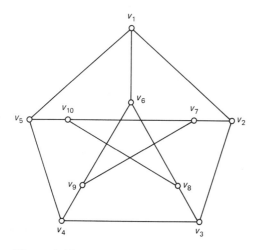

Figure 1-28. Petersen's graph: Nonplanar.

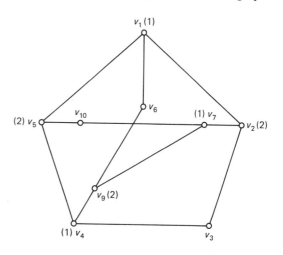

Figure 1-29. Subgraph of Petersen's graph homeomorphic to $K(3, 3)$.

G and identifying the endpoints u and v (with a single new vertex uv) so that every edge (other than (u, v)) originally incident with either u or v becomes incident with uv. We say a graph $G_1(V, E)$ is *contractible to* a graph $G_2(V, E)$ if G_2 can be obtained from G_1 by a sequence of contractions.

Theorem (Contraction Form of Kuratowski's Theorem). A graph $G(V, E)$ is planar if and only if G contains no subgraph contractible to either $K(5)$ or $K(3, 3)$.

We can use the Contraction form of Kuratowski's Theorem to directly show Petersen's graph is nonplanar, since it reduces to $K(5)$ if we contract the edges (v_i, v_{i+5}), $i = 1, \ldots, 5$.

Demoucron's planarity algorithm. We have mentioned that the Hopcroft-Tarjan planarity algorithm tests for planarity in time $O(|V|)$ and also shows how to draw a planar graph. But, instead of describing this complex algorithm, we will present the less efficient, but simpler and still polynomial, algorithm of Demoucron, et al. The algorithm is based on a criterion for when a path in a graph can be drawn through a face of a partial planar representation of the graph.

Let us define a *face* of a planar representation as a planar region bounded by edges and vertices of the representation and containing no graphical elements (vertices or edges) in its interior. The unbounded region outside the outer boundary of a planar graph is also considered a face. Let R be a planar representation of a subgraph S of a graph $G(V, E)$ and suppose we try to extend R to a planar representation of G. Then, certain constraints will immediately impose themselves.

First of all, each component of $G - R$, being a connected piece of the graph, must, in any planar extension of R, be placed within a face of R. Otherwise, if the component straddled more than one face, any path in the component connecting a pair of its vertices lying in different faces would have to cross an edge of R, contrary to the planarity of the extension. A further constraint is observed by considering those edges connecting a component c of $G - R$ to a set of vertices W in R. All the vertices in W must lie on the boundary of a single face in R. Otherwise, c would have to straddle more than one face of R, contrary to the planarity of the extension.

These constraints on how a planar representation can be extended are only necessary conditions that must be satisfied by any planar extension. Nonetheless, we will show that we can draw a planar graph merely by never violating these guidelines.

Before describing Demoucron's algorithm, we will require some additional terminology. If G is a graph and R is a planar representation of a subgraph S of G, we define a *part p of G relative to R* as either

(1) An edge (u, v) in $E(G) - E(R)$ such that u and $v \in V(R)$, or

(2) A component of $G - R$ together with the edges (called *pending edges*) that connect the component to the vertices of R.

A *contact vertex* of a part p of G relative to R is defined as a vertex of $G - R$ that is incident with a pending edge of p. We will say a planar representation R of a subgraph S of a planar graph G is *planar extendible* to G if R can be extended to a planar representation of G. The extended representation is called a *planar extension* of R to G. We will say a part p of G relative to R is *drawable* in a face f of R if there is a planar

Introduction to Graph Theory   Chap. 1

extension of R to G where p lies in f. We have already observed the following necessary condition for drawability in a face.

> Let p be a part of G relative to R.
> Then, p is drawable in a face f of R only if
> every contact vertex of p lies in $V(f)$.

The condition is certainly not sufficient for drawability, since it merely says that if the contact vertices of a part p are not all contained in the set of vertices of a face f, p cannot be drawn in f in any planar extension of R to G. For convenience, we shall say that a part p which satisfies this condition is *potentially drawable* in f, prescinding from the question of whether or not either p or G are planar.

Demoucron's algorithm works as follows. We repeatedly extend a planar representation R of a subgraph of a graph G until either R equals G or the procedure becomes blocked, in which case G is nonplanar. We start by taking R to be an arbitrary cycle from G, since a cycle is trivially planar extendible to G if G is planar. We then partition the remainder of G not yet included in R into parts of G relative to R and apply the drawability condition. For each part p of G relative to R, we identify those faces of R in which p is potentially drawable. We then extend R by selecting a part p, a face f where p is drawable, and a path q through p, and add q to R by drawing it through f in a planar manner. The process can become blocked (before G is completely represented) only if we find some part that does not satisfy the drawability condition for any face, in which case R is not planar extendible to G and G is nonplanar.

A high level procedural statement of Demoucron's algorithm follows. We rely on the previous discussion to make the meaning of the data types clear.

```
Function Draw (G, R)

(* Returns a planar representation of G in R, or fails *)

var  G: Graph
     R: Planar Representation
     p: Part
     P: Set of Part
     f: Face
     F: Set of Face
     Draw: Boolean function

Set Draw to True
Set R to some cycle in G

repeat

     Set P to the set of parts of R relative to G

     for each p in P  do  Set F(p) to the set of faces of R
                             in which p is drawable

     if       F(p) is empty for any p
```

    then   **Set** Draw to False

    else if   For some part p, F(p) contains only a single face

        then Let f be the (unique) face in which p is drawable

        else Let p be any part and let f be a face in F(p)

        Let q be a path between a pair of contact vertices
        of p with R that contains only edges of p

        Set R to R ∪ q

    until   R = G **or** **not** Draw
    End_Function_Draw

       The construction of a path q between a pair of contact vertices c_1 and c_2 on the boundary of a face f of a planar representation R is illustrated in Figure 1-30. An application of Demoucron's algorithm to Petersen's graph is shown in Figures 1-31 and 1-32. Initially, R consists of the cycle v_1-v_2-v_3-v_4-v_5-v_1. There is only one part of G relative to R, consisting of all the remaining vertices and edges of G, that is, the induced subgraph on vertices $v_6..v_{10}$ and the edges connecting this subgraph to the cycle R. The contact vertices of this part are $v_1..v_5$. Any path through this unique part which connects one of its contact vertices to another can serve as the path q described in the procedure. We will select v_1-v_6-v_8-v_3 as q and update R to $R \cup q$. The new subgraph R has two interior faces and a single exterior face. There is only one part of G relative to R, as shown in Figure 1-32, and its contact vertices are $\{v_2, v_4, v_5, v_6, v_8\}$. Since none of the faces of R contains this set of contact vertices, the part is not drawable in any face; hence we conclude Petersen's graph is nonplanar.

      Theorem (Correctness of Demoucron's Algorithm).   Let $G(V, E)$ be a graph. Then, Demoucron's algorithm finds a planar embedding of G if G is planar or correctly recognizes G as nonplanar.

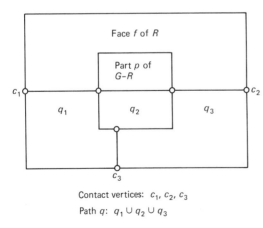

Contact vertices: c_1, c_2, c_3
Path q: $q_1 \cup q_2 \cup q_3$

Figure 1-30. Addition of path q to R.

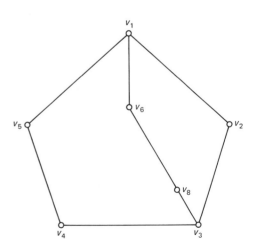

Figure 1-31. Second R construct for Petersen's graph.

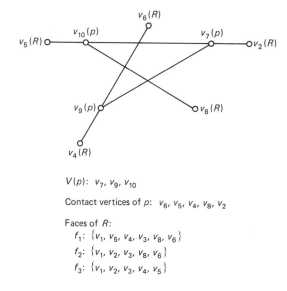

$V(p)$: v_7, v_9, v_{10}

Contact vertices of p: v_6, v_5, v_4, v_8, v_2

Faces of R:
  f_1: $\{v_1, v_5, v_4, v_3, v_8, v_6\}$
  f_2: $\{v_1, v_2, v_3, v_8, v_6\}$
  f_3: $\{v_1, v_2, v_3, v_4, v_5\}$

Figure 1-32. A unique part p relative to R.

The proof of the theorem is as follows. We use an induction over the (implicit) loop index of the algorithm. We show that if G is planar, the representation R constructed by the algorithm is always planar extendible to a planar representation of G. The initial representation R is a cycle, and so is trivially planar extendible, provided G is planar in the first place. In general, we show that if the current representation R is planar extendible to G, its updated representation remains planar extendible. If R is planar extendible, every part of G relative to R must have some face of R in which it is drawable. We will distinguish two cases depending on whether or not some part is drawable in only a single face.

If a part p has a unique face f in which it is drawable, since R is planar extendible and there is only one possible choice of face in which to embed p with respect to R, then p must lie in f in every possible planar extension of R to G. Consequently, we may draw the path q through f and the resulting extension of R remains planar extendible.

Suppose, on the other hand, that for every part p of G relative to R there are at least two faces in which p is drawable. We will show that in this case the choice of face in which p is drawn is not critical. Let f be a face in which p is drawable. Then, we show there exists a planar extension of R to G for which p lies in f. Let RX denote an extension of R to G where p lies in a face f' not equal to f. Then, the contact vertices of p must lie on both the border of f and the border of f', and so must lie on the shared boundary B between f and f'. Consider all the parts of G relative to R whose contact vertices lie on B. Some of these parts lie in f in RX and some lie in f' in RX. Define a new planar representation RX' of G by flipping these parts about B. That is, flip those parts lying in f in RX so they lie in f' in RX', and flip those parts lying in f' in RX so they lie in f in RX'. Then, RX' is a planar representation of G for which p lies in f. Therefore, we can draw p in f and R remains planar extendible, as required.

Thus far, we have shown that if G is planar, the algorithm can extend a planar representation of a subgraph until all of G is drawn. On the other hand, of course, if G is nonplanar, the algorithm will recognize this at some stage by finding a part with no face in which it can be drawn, which completes the proof of the theorem.

Planar graph theory. We will now prove some elementary but basic results of planar graph theory. The first theorem is simple and was one of the first contributions to the subject. Its name derives from the relation it identifies between the number of vertices, edges, and faces of a polyhedron (cube, tetrahedron, etc.). The edges of these solids define planar graphs when they are appropriately projected on the plane.

Theorem (Euler Polyhedron Formula). Let $G(V, E)$ be a connected planar graph and let $|F|$ denote the number of faces in a planar embedding of G. Then,

$$|F| + |V| = |E| + 2.$$

See Figure 1-33 for an example.

The proof of the theorem is as follows. It is by induction on the number of edges of the graph. If the graph has only one edge, $|V| = 2$, $|E| = 1$, and $|F| = 1$; so the theorem is trivial. By induction, we assume the theorem is true for any graph with n edges, and then consider what happens if we add a further edge. (Refer to Figure 1-34.) If the additional edge leads to a new vertex, $|V|$ and $|E|$ will both increase by 1, while $|F|$ will remain unchanged; so the balance of the equation will be maintained. On the other hand, if the new edge is between existing vertices, the edge must divide an existing face into two parts, thus increasing both $|F|$ and $|E|$ by 1, which again preserves the equation. This completes the proof.

Theorem (Linear Bound on $|E|$). If $G(V, E)$ is a planar graph, then

$$|E| \leq 3|V| - 6.$$

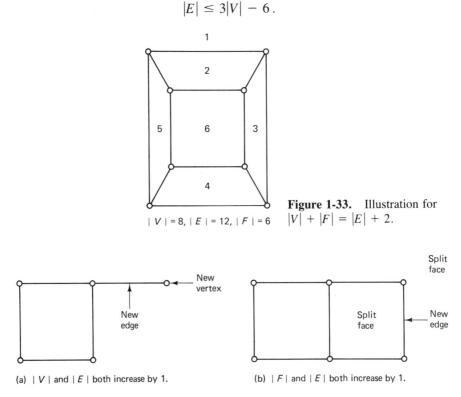

$|V| = 8, |E| = 12, |F| = 6$

Figure 1-33. Illustration for $|V| + |F| = |E| + 2$.

New vertex

New edge

Split face

Split face

New edge

(a) $|V|$ and $|E|$ both increase by 1.

(b) $|F|$ and $|E|$ both increase by 1.

Figure 1-34. The different effects of adding an edge.

The proof is as follows. We let $|F_i|$ denote the number of edges bounding the ith face of some planar representation of G. Each face must contain at least 3 bounding edges. Therefore,

$$\sum_{i=1}^{|F|} |F_i| \geq 3|F|.$$

Since each edge borders exactly two faces, the summation counts each edge twice. Therefore,

$$\sum_{i=1}^{|F|} |F_i| = 2|E|.$$

Combining these results and using the Euler Polyhedron Formula, we obtain

$$|E| \leq 3|V| - 6,$$

as was to be shown.

Theorem (Nonplanarity of Kuratowski Graphs). The graphs $K(3, 3)$ and $K(5)$ are nonplanar.

The proof of this theorem is as follows. The nonplanarity of $K(5)$ follows by contradiction from the Linear Bound Theorem, since if $K(5)$ were planar we would have $|E| (= 10) \leq 3|V| - 6 = 9$, which is a contradiction. For $K(3, 3)$, we argue as follows. By Euler's Theorem, $|F| = |E| - |V| + 2 = 5$. However, since every cycle in $K(3, 3)$ has at least four edges, we can refine the proof of the Linear Bound Theorem in this case to prove that $|F| \leq 4$. That is, if $K(3, 3)$ were planar we would have

$$\sum_{i=1}^{|F|} |F_i| \geq 4|F|.$$

whence, emulating the Linear Bound proof, we could conclude that $|E| (= 9) \geq 2|F|$, so that $|F| \leq 4$, contrary to the previous lower bound of 5 for $|F|$. Therefore, $K(3, 3)$ must be nonplanar. This completes the proof of the theorem.

The following upper bound on the minimum degree will prove useful when we consider an algorithm for five-coloring a planar graph in Chapter 7.

Theorem (Minimum Degree Bound). If $G(V, E)$ is a planar graph, $min(G) \leq 5$.

The proof is by contradiction. Suppose that the minimum degree of G is at least 6. Then,

$$\sum_{i=1}^{|V|} \deg(v_i) \geq 6|V|.$$

Since the sum of the degrees is always $2|E|$, it follows that $|E|$ is at least $3|V|$, contrary to the Linear Bound Theorem. It follows that the minimum degree of G must be at most 5, which completes the proof.

Embedding graphs on manifolds. The attempted embedding of $K(3, 3)$ in Figure 1-26 succeeds except for the single edge x. But, $K(3, 3)$ can be embedded on the surface of a *torus,* the closed two-dimensional manifold that corresponds to the surface of a doughnut. We can represent a torus in a planar fashion by cutting it across its tubu-

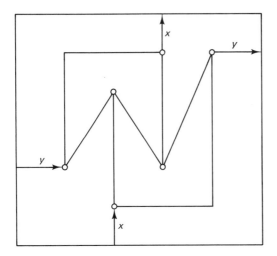

Figure 1-35. Embedding of $K(3,3)$ on torus.

lar cross-section, and then cutting the resulting cylinder to form a rectangle. We consider the opposite borders of the resulting rectangle as identified. Figure 1-35 gives a toroidal embedding of $K(3,3)$. We think of the edge x in the figure, that strikes out vertically towards the upper border of the rectangle, as passing around the back of the torus, reappearing on the lower border after having circumnavigated the tubular dimension of the torus. The edge y that starts out to the right and reappears at the opposite point on the left border of the rectangle corresponds to a radial circumnavigation of the torus.

This example suggests two common generalizations of the concept of planar embedding: The genus of a graph and the crossing number of a graph.

The *genus of a* (closed two-dimensional) *surface* is the number of handles on the surface. Thus, we can consider a sphere as having genus zero, while a torus, which can be visualized as a sphere with one handle has genus one. The *genus of a graph G* is then defined as the genus of the surface of least genus on which G can be embedded, that is, drawn without spurious edge intersections. For example, since we can draw $K(3,3)$ on the torus but not on the sphere (which is equivalent for present purposes to the plane), $K(3,3)$ has genus one. $K(5)$, $K(6)$, and $K(7)$ can also be embedded on the torus; so their genus' are also one. On the other hand, $K(8)$ requires a two handled sphere for embedding and so has genus two.

The *crossing number* of a graph $G(V,E)$ is defined as the minimum number of edge crossings among all possible embeddings of G in the plane. Obviously, the crossing number of a planar graph equals zero while, as we have just seen, the crossing number of $K(3,3)$ equals one.

1-7 EULERIAN GRAPHS

An *euler trail* in a graph $G(V,E)$ is a closed trail (or circuit) that spans all the edges of G. An *open euler trail* is an open trail that spans all the edges of G. An *eulerian graph* is a graph that contains an eulerian trail. A *semieulerian graph* is a graph that contains an open euler trail. Eulerian, semieulerian, and noneulerian graphs are shown in Figure 1-36.

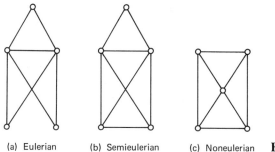

(a) Eulerian   (b) Semieulerian   (c) Noneulerian   **Figure 1-36.**   Example graphs.

For convenience, we will use the following notation (for this section only). Let T be a trail (open or closed) in a connected graph $G(V, E)$. Then, we denote by $G - T$ the graph obtained by removing $E(T)$ from $E(G)$ along with any vertices in $V(G)$ isolated as a result of the removal of $E(T)$ from G. Eulerian graphs have a simple characterization in terms of their degree sequences.

Theorem (Characterization of Eulerian Graphs). Let $G(V, E)$ be a connected graph. Then, G is eulerian if and only if the degree of every vertex in G is even.

The proof of the theorem is as follows. The necessity of the condition is obvious. If G is eulerian, the euler trail clearly induces a graph whose vertices have even degree. We can prove sufficiency as follows. Since every vertex in G has even degree, G must contain some closed trail T. Since G is connected, every component of $G - T$ must be incident with some vertex of T. The subgraphs induced by each of these components are connected and have degree sequences with even degrees; so we may assume by induction that each of these components is eulerian. Let T' be an euler trail in a component that intersects T at a vertex v. We can combine T and T' into one euler trail by starting T at any of its vertices, proceeding until we reach v, and then following the trail T' until we have traversed it completely; whence we continue with the traversal of T. If we repeat this process, we eventually obtain a closed spanning trail of G. This completes the proof of the theorem.

If a connected graph has only two vertices of odd degree, it is semieulerian. This is a special case of the following more general theorem.

Theorem (Euler Trail Decomposition). Let $G(V, E)$ be a connected graph with $2k$ vertices of odd degree. Then, the edges of G can be partitioned into k edge disjoint open trails $E_1, \ldots, E_k$.

The proof of the theorem is as follows. Denote the $2k$ vertices of odd degree by $v_1, \ldots, v_k$ and $w_1, \ldots, w_k$. Add k new vertices $u_1, \ldots, u_k$ to G together with $2k$ new edges (v_i, u_i) and (u_i, w_i), where $i = 1, \ldots, k$. Denote the resulting graph by G'. By the previous theorem, G' is eulerian, and so has an euler trail. Since every new vertex u_i is of degree two in G', removing these vertices from the trail, breaks the euler trail on G' into k edge disjoint open tails, which together cover the edges of G. This completes the proof.

An *eulerian directed graph* is a digraph containing a directed euler trail. We say a digraph G is connected if the underlying undirected graph of G is connected. The following theorem is easily proved.

Theorem (Characterization of Directed Eulerian Graphs). Let $G(V, E)$ be a connected digraph. Then, G is eulerian if and only if for every vertex v in $V(G)$, indeg(v) equals outdeg(v).

Euler trail algorithms. We will describe two algorithms for finding euler trails in graphs. The first algorithm, which is due to Fleury, has an intuitive appeal; while the second algorithm, which is motivated by the proof of the Eulerian Characterization Theorem, is more efficient.

Fleury's algorithm is a visually convenient way of constructing an euler trail in an eulerian graph but suffers from $O(|E|^2)$ performance. The idea of Fleury's algorithm is to successively trace out the edges of a trail, erasing traversed edges and any resulting isolated vertices as we proceed, never traversing an edge if in doing so we would disconnect the remaining graph into nontrivial components or isolate the starting vertex before all the edges are traversed. Thus, Fleury's algorithm repeatedly extends an incomplete eulerian trail T from its terminal vertex v by appending to T any edge x incident with v which is not a bridge of $G - T$ and whose removal would not isolate the starting vertex before all the edges are traversed. Since it takes $O(|E|)$ time to test whether an edge is a bridge, each edge only has to be tested once, and there are $|E|$ edges in G, Fleury's algorithm has performance $O(|E|^2)$.

Refer to Figure 1-37 for an illustration of the pitfalls avoided by Fleury's algorithm. If we start a trail at the vertex c traversing it until we obtain $T = c\text{-}b\text{-}a$, at that point the edge (a, d) which is incident with the endpoint a of T is a bridge of $G - T$. If we follow the edge selection principle in Fleury's algorithm and continue the trail via either of the edges (a, e) or (a, g), we will be able to successfully complete an euler trail. On the other hand, if we extend T via the edge (a, d), which is contrary to the bridge avoidance criterion in Fleury, the resulting trail becomes trapped when it reaches c, before it has traversed all of $E(G)$. The following theorem is easily proved.

Theorem (Fleury's Algorithm). Let $G(V, E)$ be an eulerian graph. Then, Fleury's algorithm finds an euler trail in G in $O(|E|^2)$ time.

We will now describe an $O(|E|)$ algorithm for euler trails based on the proof of the eulerian characterization theorem. We start with a trail T consisting of a single vertex and extend it until it becomes blocked. We then select a previously reached but in-

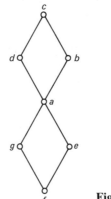

Figure 1-37. Fleury example.

completely traversed vertex from which we initiate a new closed trail T'. We then subsequently patch T' onto T, forming a new and more extensive trail. The process continues until all of $E(G)$ is traversed.

If we denote the list of vertices that have been reached by the partial euler trail T but that still have untraversed incident edges by PL (meaning the Pending List), the vertices on PL correspond to those vertices in the characterization proof at which additional closed trails were patched onto an existing trail to produce a more extensive closed trail. The pending list becomes exhausted precisely when we have completed an euler trail on the component of G in which the trail was initiated, which in the case that G is connected is an euler trail on G.

We will now describe the procedure for constructing the euler trail more formally. We first consider the design of the requisite data structures. We will represent the graph $G(V, E)$ as a linear array with doubly linked adjacency lists and with shared representatives for its edges to facilitate the edge deletions that will be required by the algorithm. We include a field, denoted PL-Pointer(v), at each vertex v, which points to the entry for v, if any, on the pending list PL. Another direct access pointer, denoted by Occurrence(v), points to an entry for v on T, if there is one. Both T and PL are represented by doubly linked lists. This organization, in conjunction with Occurrence, allows us to efficiently patch additional closed trails onto T. Similarly, in conjunction with the PL-Pointer field, it allows us to easily delete elements from PL. The corresponding type definitions are as follows.

```
type  Graph = record
                H(|V(N)|): Vertex
              end

      Vertex = record
                 Positional-Pointer, Successor: Edge pointer
                 PL-Pointer, Occurrence: Entry pointer
               end

      Edge  = record
                Shared-rep: Shared-edge pointer
                Edge-successor,
                    Edge-predecessor: Edge pointer
              end

      Shared-edge =
                record
                  E(1,2): 1..|V|
                  Endpoints(2): Edge pointer
                end

type  List  = record
                Head, Tail: Entry pointer
              end

      Entry = record
                Index: 1..|V|
                Predecessor, Successor: Entry pointer
              end
```

The procedure for constructing the euler trail is named Euler. Euler uses several subprocedures, which we will now describe. Create(T) initializes a doubly linked list T. Put(u,T) adds a vertex u to the end of T; while Put(u,PL,G) adds u to the pending list PL (it has a null effect if u is already on the list) and sets the corresponding PL-Pointer in G to point to that entry. Get(PL,w) returns an element w from the pending list PL, without deleting it, and fails if there is none. The function Next(G,w,v) returns in v the next neighbor of w or fails if there is none. Delete_edge(G,w,v) deletes the edge (w, v) from G. Empty(G,u) succeeds if the adjacency list at u is empty; Single(G,u) succeeds if the adjacency list at u has at most one more element; while Empty(G) succeeds if G is empty. Delete(w,G,PL) deletes the entry for w on PL using PL-Pointer(w) in G (and has a null effect if w is not on the list).

The procedure Patch(T,T',w) patches the closed trail T' that begins and ends at w onto the trail T by replacing one occurrence of w on T by the trail T'. Patch has a simple string substitution interpretation.

(1) If the trail T is

$$x_1\text{-}x_2\text{-}\ldots\text{-}a\text{-}(w)\text{-}b\text{-}\ldots\text{-}x_n,$$

(2) and the substitute trail T' is

$$(w\text{-}w_1\text{-}\ldots\text{-}w_k\text{-}w),$$

(3) the new trail $T \cup T'$ after substitution is

$$x_1\text{-}x_2\text{-}\ldots\text{-}a\text{-}(w\text{-}w_1\text{-}\ldots\text{-}w_k\text{-}w)\text{-}b\text{-}\ldots\text{-}x_n, \quad \text{or}$$

$$x_1\text{-}x_2\text{-}\ldots\text{-}a\text{-}w\text{-}w_1\text{-}\ldots\text{-}w_k\text{-}w\text{-}b\text{-}\ldots\text{-}x_n.$$

The doubly linked representations for T and T' together with the Occurrence field in G which points to an occurrence of w in T allow an $O(1)$ realization of Patch. The procedure for Euler follows.

```
Function Euler (G, u, T)

(* Returns euler trail for G starting at u in T, or fails *)

var G: Graph
    u,w,v: 1..|V|
    T, T', PL: List
    Next, Get, Empty, Single, Euler: Boolean function

Create(T); Put(u,T); Create(PL); Put(u,PL,G)

while  Get(PL,w)  do

    Create(T'); Put(w,T')
    while Next(G,w,v)  do   Put(v,T')
                            Delete_edge (G,w,v)
                            if Empty(G,w) then Delete(w,G,PL)
                            if Single(G,v)  then Delete(v,G,PL)
                                            else Put(v,PL,G)
                            Set w to v
    Patch(T,T',w)

Set Euler to Empty (G)
End_Function_Euler
```

1-8 HAMILTONIAN GRAPHS

We have already defined hamiltonian graphs as graphs that contain spanning cycles. There is an extensive and profound theory of hamiltonian graphs, which is currently largely of theoretical interest. We will consider this topic here partly because, like Mount Everest, "It is There," but also because the problem of recognizing such graphs plays a paradigmatic role in algorithmic complexity theory. Indeed, the recognition of these graphs is a classic instance of a problem for which straightforward search algorithms are available but for which no efficient algorithm seems possible.

We call a spanning cycle in a hamiltonian graph a *hamiltonian cycle*, while a spanning path in a graph is called a *hamiltonian path*. Analogous terms are defined for digraphs. The problem of determining whether a graph has a hamiltonian cycle is called the *hamiltonian cycle problem*. The related *traveling salesman problem* seeks to find the shortest spanning circuit in a weighted graph, which, when the edge weights satisfy the triangle inequality, can be shown to be a hamiltonian cycle. The hamiltonian cycle and traveling salesman problems are both NP-Complete, that is, as we shall see in Chapter 10, they are probably unsolvable in polynomial time. We will describe a backtracking algorithm for finding hamiltonian cycles, a search technique often appropriate for nonpolynomial problems. We will also describe an approximation algorithm for the euclidean version of the traveling salesman problem.

Model: Optimal schedules. The following problem can be solved by finding a shortest hamiltonian path in a weighted digraph. Let $\{P_i, i = 1, \ldots, n\}$ be a set of n processes, all of which need a resource which they access sequentially. An ordered list of the processes constitutes a process schedule. If a process P_i is scheduled just prior to process P_j, a reset cost c_{ij} is incurred in preparing the resource to run P_j. The cost of a schedule is defined as the total of the reset costs summed over all the whole schedule. We can model the scheduling of the set of processes and the reset costs as a weighted digraph G, where the processes correspond to the vertices of the digraph and the weight of an edge (i, j) equals the reset cost c_{ij}. The cost of a process schedule then corresponds to the total cost of the edges between successive processes of the schedule. If we define an optimal schedule as a schedule of minimum cost, an optimal schedule corresponds to a shortest hamiltonian path in G.

Basic concepts. In general, it is difficult to determine if a graph has a hamiltonian cycle or hamiltonian path, but there are some simple necessary and/or sufficient conditions for a graph to be hamiltonian. The following theorem of Ore is well-known.

Theorem (Ore's Path and Cycle Length Condition). Let $G(V, E)$ be a graph of order $|V| \geq 3$, and suppose that for every pair of nonadjacent vertices u and v in G

$$\deg(u) + \deg(v) \geq M,$$

for an integer M. If M equals $|V|$, G is hamiltonian, while if G is connected, it contains a path of length M.

We shall prove the theorem in the special case that M equals $|V|$ and leave the generalization to the complete theorem to the reader. The proof is by contradiction. Suppose the theorem does not hold for some order $|V|$. Let G be a nonhamiltonian graph of order $|V| = p$ which satisfies the conditions of the theorem and has maximum size among all such graphs. Let u and v be nonadjacent vertices in G. Then, $G \cup (u, v)$

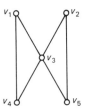

Figure 1-38. Sharpness of Ore Hamiltonian condition.

must be hamiltonian, and so there must be a hamiltonian path P in G from u to v. Let the successive vertices of P be $u = u_1, u_2, \ldots, u_p = v$. If u is adjacent to u_i, then v cannot be adjacent to u_{i-1}. Otherwise, the sequence $u_1, u_2, \ldots, u_{i-1}, v, u_{p-1}, \ldots, u_i, u_1$ would determine a hamiltonian cycle in G, contrary to assumption. Therefore, for every vertex in G that u is adjacent to, there is a vertex in G that v is not adjacent to. That is, $\deg(v) \le |V| - 1 - \deg(u)$, contrary to the condition of the theorem. It follows that G must be hamiltonian, which completes the proof.

We refer to Figure 1-38 for an example illustrating the sharpness of the theorem. The conclusions of the theorem can be strengthened, for when $M = |V|$ the graph can be shown to have cycles of every order, unless $|V|$ is even and G is isomorphic to $K(|V|/2, |V|/2)$.

The Theorem of Ore provides a sufficient condition for the existence of a hamiltonian cycle. The following theorem gives a necessary and sufficient condition. First, we define the *closure* of a graph $G(V, E)$ of order $|V|$ as the graph obtained from G by recursively inserting an edge between every pair of nonadjacent vertices of degree sum at least $|V|$ (or which become so recursively as the result of adding edges).

Theorem (Closure Condition for Hamiltonicity). A graph $G(V, E)$ is hamiltonian if and only if the closure of G is hamiltonian.

We refer to Figure 1-39 for an example of the closure operation. The final graph is $K(6)$, which is hamiltonian; so the original graph is hamiltonian.

There is a simple condition for a digraph to have a hamiltonian path.

Theorem (Redei's Condition). If $G(V, E)$ is a digraph whose underlying graph is complete, G has a directed hamiltonian path.

The proof is by an induction on the order of G. Assume the theorem is true for digraphs with order at most p, and consider a digraph G of order $p + 1$. Let u be any vertex in G. Then, by induction, the digraph $G - u$ has a hamiltonian path P: $u_1, \ldots, u_p$. By assumption, either (u, u_1) or (u_1, u) is in G. If (u, u_1) is in G, then u, $u_1, \ldots, u_p$ determines a hamiltonian path in G, as required. If (u_1, u) is in G, let u_i be the first vertex on P for which (u, u_i) is in G, if any. If u_i exists, then $u_1, u_2, \ldots, u_{i-1}$, $u, u_i, u_{i+1}, \ldots, u_p$ determines a hamiltonian path in G. Otherwise, $u_1, u_2, \ldots, u_p, u$ determines a hamiltonian path in G. This completes the proof.

There are interesting relations between hamiltonicity, connectivity, planarity, and the powers of a graph. We call a graph $G(V, E)$ *hamiltonian connected* if there is a hamiltonian path between every pair of vertices in G. A hamiltonian graph is necessarily hamiltonian connected, though not vice versa. We have the following theorem.

Theorem (Hamiltonian Powers). If $G(V, E)$ is one connected, the cube of G is hamiltonian connected. If $G(V, E)$ is two connected, the square of G is hamiltonian connected.

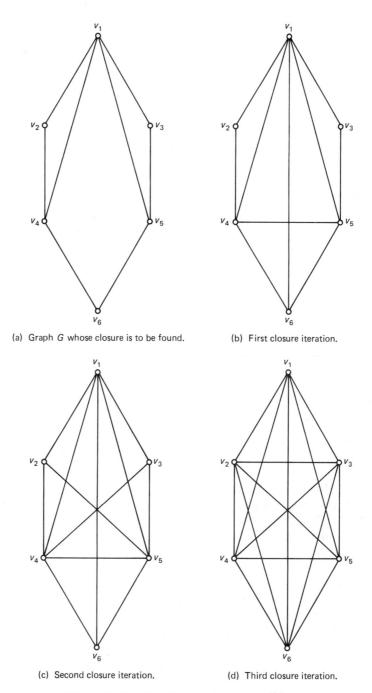

(a) Graph G whose closure is to be found.

(b) First closure iteration.

(c) Second closure iteration.

(d) Third closure iteration.

Figure 1-39. Hamiltonian closure condition.

The square of the one connected graph in Figure 1-40 is nonhamiltonian, proving the sharpness of the first condition in the theorem. Contrary to the progression suggested by the theorem, it is not the case that if G is three connected, G must be hamil-

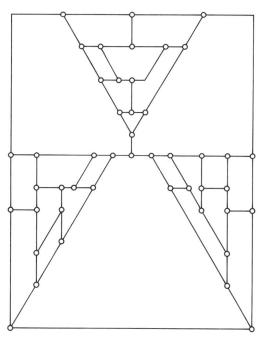

Figure 1-41. Cubic planar three-connected but nonhamiltonian.

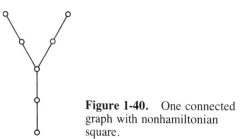

Figure 1-40. One connected graph with nonhamiltonian square.

tonian. Indeed, Figure 1-41 shows Tutte's famous counterexample to the longstanding conjecture that even three connected cubic planar graphs were necessarily hamiltonian.

Theorem (Tutte's Condition). If $G(V, E)$ is four connected and planar, G is hamiltonian.

Backtracking algorithm. Backtracking is a general technique for generating solutions to a combinatorial problem by systematically extending partial solutions to the problem. The technique is described in greater detail in Chapters 2 and 8. We will use it here to find hamiltonian cycles in a graph G. We will assume G is represented by the type adjacency matrix (with matrix **Adj**), described in Section 1-2.

The algorithm is stated in Find_Hamiltonian_Path(G). This finds all the hamiltonian paths in a graph $G(V, E)$ by repeatedly extending partial hamiltonian paths. The successive vertices of the path are stored in an array **Path**. Find_Hamiltonian_Path provides the driver logic for the backtracking algorithm, while the procedure Next(Path,k) does the work of getting the next candidate vertex for extending the path. Next returns, in Path(k), the next vertex which is adjacent to the current endpoint of the path, Path(k − 1), and which has an index higher than the current value of Path(k). If there is no such vertex, Next returns the dummy index $|V(G)| + 1$. The new path is **Path**(1), ..., **Path**(k − 1) | **Path**(k), where | denotes path concatenation. We use an array **Status** to indicate if a vertex is on the current path. Status(i) is 1 if vertex i is on the path, and 0 otherwise. To simplify the implementation of Next, we use the conditional operator or*, which leaves a later operand unevaluated if an earlier one is true.

Procedure Find_Hamiltonian_Paths (G)

(* Finds all hamiltonian Paths in G *)

var G: Graph
    k: 1..|V|
    Path(1..|V|): 0..|V| + 1
    Status(1..|V| + 1): 0..1
    No_More_Paths: Boolean

Set Path (1..|V|) to 0
Set Status (1..|V| + 1) to 0
Set k to 1
Set No_More_Paths to False

repeat

    Next(Path, k)

    case

    1: $k < |V|$ **and** Path(k) $\leq$ |V| : **Set** k to k + 1

    2: $k = |V|$ **and** Path(k) $\leq$ |V| : Display Path(1..|V|)
                                             Set Status(Path(|V|)) to 0

    3: $k > 1$ **and** Path(k) > |V| : **Set** Path(k) to 0
                                              Set Status(Path(k − 1)) to 0
                                              Set k to k − 1

    4: $k = 1$ **and** Path(k) > |V| : **Set** No-More_Paths to True

until   No-More_Paths

End_Procedure_Find_Hamiltonian_Paths

Procedure Next(Path,k)

(* Returns in Path(k) the next vertex adjacent to Path(k − 1)
   not currently on the path *)

var G: Graph
    k: 1..|V|
    Path(1..|V|): 0..|V| + 1
    Status(1..|V| + 1): 0..1

repeat

  repeat Increment Path(k)   **until** Status(Path(k)) = 0

until k = 1 **or*** Path(k) > |V| **or*** Adj(Path(k),Path(k-1)) = 1

if Path(k) $\leq$ |V| **then**   **Set** Status(Path(k)) to 1

End_Procedure_Next

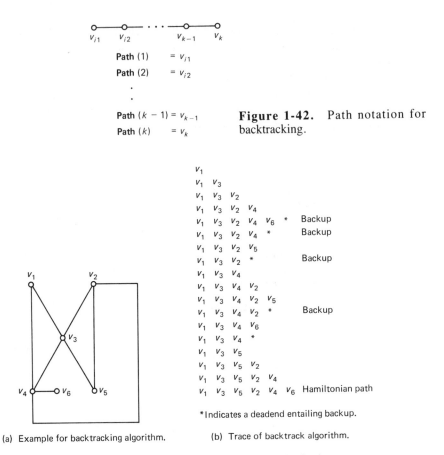

Path (1) = v_{i1}
Path (2) = v_{i2}
·
·
Path $(k-1)$ = v_{k-1}
Path (k) = v_k

Figure 1-42. Path notation for backtracking.

v_1
v_1 v_3
v_1 v_3 v_2
v_1 v_3 v_2 v_4
v_1 v_3 v_2 v_4 v_6 * Backup
v_1 v_3 v_2 v_4 * Backup
v_1 v_3 v_2 v_5
v_1 v_3 v_2 * Backup
v_1 v_3 v_4
v_1 v_3 v_4 v_2
v_1 v_3 v_4 v_2 v_5
v_1 v_3 v_4 v_2 * Backup
v_1 v_3 v_4 v_6
v_1 v_3 v_4 *
v_1 v_3 v_5
v_1 v_3 v_5 v_2
v_1 v_3 v_5 v_2 v_4
v_1 v_3 v_5 v_2 v_4 v_6 Hamiltonian path

*Indicates a deadend entailing backup.

(a) Example for backtracking algorithm.   (b) Trace of backtrack algorithm.

Figure 1-43. Example for Find_Hamiltonian_Path.

We refer to Figures 1-42 and 1-43 for an example.

We can adapt the same procedure with minor variations to find a variety of other combinatorial objects such as

(1) Hamiltonian cycles,
(2) A longest path starting at a given vertex,
(3) A longest path between a given pair of vertices,
(4) A longest path between a given pair of vertices and which avoids a given set of vertices, and
(5) Paths of length at least k starting at a given vertex.

Approximate algorithm. Let $G(V, E)$ be a complete graph with positive weights assigned to each edge. We denote the weight assigned to edge (i, j) by $c(i, j)$. We will say G is a *euclidean graph* if the edge weights satisfy the *triangle inequality*

$$c(i, j) + c(j, k) \geq c(i, k),$$

for every three vertices i, j, and k. For euclidean graphs, optimal circuits are optimal cycles.

Theorem (Euclidean Hamiltonian Cycles). If $G(V, E)$ is a euclidean graph, a hamiltonian cycle on G of least cost is also a solution to the traveling salesman problem on G.

The proof of the theorem is as follows. We will show that a least cost circuit can always be transformed into a hamiltonian cycle of equal cost. Thus, suppose C is a least cost circuit in G which is not a cycle. Then, C must contain at least one multiply covered vertex u and so can be represented as a sequence

$$v_1 - \ldots - u - \ldots - v_i - u - v_j - \ldots - v_k - v_1 .$$

If we replace the edges (v_i, u) and (u, v_j) by the single edge (v_i, v_j), then u still lies on the modified cycle since u had already been spanned before this occurrence of u. It follows from the triangle inequality that

$$c(v_i, v_j) \leq c(v_i, u) + c(u, v_j) .$$

Therefore, replacement does not increase the cost of the circuit and the new circuit remains closed and spanning. If we repeat this procedure for every such vertex u that appears at multiple points on the circuit, we must eventually obtain a simple, non-self-intersecting circuit, that is, a cycle. By construction, the cycle is spanning and of minimum cost. This completes the proof of the theorem.

Although it is computationally difficult to obtain an exact solution to the traveling salesman problem, we can easily obtain a good approximation to an optimal solution. Indeed, if the graph is euclidean, by the preceding theorem we can even convert an approximately optimal circuit to an approximately optimal hamiltonian cycle. We will assume that G is euclidean.

The procedure for an approximate solution to the traveling salesman problem on G is as follows:

(1) Construct a minimum weight spanning tree *mst* on G,

(2) Convert *mst* to an eulerian graph *eul* by doubling its edges,

(3) Construct an euler trail *et* on *eul*, and

(4) Apply the euclidean transformation to convert *eul* to a spanning cycle *cyc*.

The minimum weight spanning tree referred to in (1) can be efficiently constructed using techniques described in Chapter 4. The doubling of the edges in (2) transforms the graph into what is strictly speaking a multi-graph (where parallel edges are allowed). Nonetheless, the standard results for eulerian graphs developed in the preceding section still apply, whence *eul* is eulerian and so contains an euler trail *et*. Then, in (4), we apply the circuit-to-cycle transformation described previously. We refer to Figure 1-44 for an example of the procedure. In the example, the approximately optimal solution C_3 is actually optimal. It is easy to establish the following bound on the error of the approximation.

Theorem (TSP Error Bound). Let $G(V, E)$ be a euclidean graph. Let S denote an optimal solution to the traveling salesman problem on G and let cost(S) denote its weight. Let *cyc* denote the spanning cycle constructed by the approximation procedure we have described and let cost(cyc) denote its weight. Then,

$$\text{cost}(cyc) \leq 2 \, \text{cost}(S) .$$

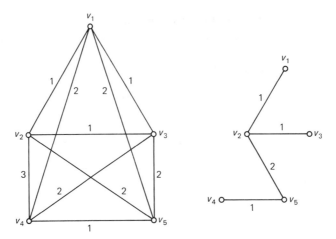

(a) Graph G with euclidean weights.   (b) Minimum spanning tree for G.

								Length
C_0:	v_1	v_2	$(v_3$	v_2	$v_5)$	v_4	v_5 v_2 v_1	10
C_1:	v_1	v_2	v_3		v_5	$(v_4$	v_5 $v_2)$ v_1	9
C_2:	v_1	v_2	v_3		v_5	$(v_4$	v_2 $v_1)$	9
C_3:	v_1	v_2	v_3		v_5	v_4	v_1	7

(c) Successive euclidean reductions of initial circuit C_0.

Figure 1-44.   Euclidean TSP approximation example.

The proof is simple. The total weight of the spanning tree *mst* is certainly less than or equal to the weight of an optimal circuit S since any spanning circuit contains a spanning tree. Therefore, the weight of *et* is at most twice the weight of *mst*. It follows that the cost of the spanning cycle *cyc* is at most twice the cost of an optimal solution.

REFERENCES AND FURTHER READING

Behzad, et al. (1979) is an excellent, detailed introduction to pure graph theory. Wilson (1985) is a delightful briefer introduction to the subject. Harary (1971) is classical and comprehensive. See Capobianco and Molluzzo (1978) for a comprehensive collection of examples and counterexamples to conjectures in graph theory. Roberts (1984) has many interesting examples of the applications of combinatorics in general, and graphs in particular, as does the older Busacker and Saaty (1965).

Gibbons (1985) is a very readable introduction to graph algorithms. Reingold, Nievergelt and Deo (1977), Christofides (1975), and Papadimitriou and Steiglitz (1982) cover combinatorial and graph algorithms, as do Gondran and Minoux (1984) and Swamy and Thulasiraman (1981). Tarjan (1983) has extensive discussions of efficient data structures for graph algorithms. For data structures, see Standish (1980) and all the volumes of Knuth. McHugh (1986) gives an overview of data structures. Welsh (1983) reviews random combinatorial algorithms and refers to Lovasz (1979). See also, Lovasz and Plummer (1987) and the references given for matching in Chapter 8. See

Bondy and Murty (1976) for the algorithm of Demoucron, et al., as well as Demoucron, Malgrange, and Pertuiset (1964); also see Gibbons (1985) for a discussion. The procedure for drawing a planar graph with straight line segments described in the exercises is from Tutte (1963). The linear time planarity algorithm was given in Hopcroft and Tarjan (1974).

EXERCISES

1. Let $G(V, E)$ be a graph. Prove either G or G^c is connected.

2. What is the largest number of edges in a disconnected graph G of order $|V|$?

3. Show a graph of order $|V|$ with more than $|V|^2/4$ edges cannot be bipartite.

4. Construct two nonisomorphic graphs with the same reachability matrix.

5. What are the nonisomorphic graphs of order 5?

6. Show G is bipartite if and only if its adjacency matrix can be rearranged and partitioned into submatrices $\mathbf{A}_{ij}$ ($i, j = 1, 2$), where $\mathbf{A}_{11}$ and $\mathbf{A}_{22}$ are 0, and $\mathbf{A}_{12}$ and $\mathbf{A}_{21}$ are transposes. Show G is disconnected if and only if its adjacency matrix can be rearranged and partitioned into submatrices so that $\mathbf{A}_{12}$ and $\mathbf{A}_{21}$ are 0. What partitions are feasible if G has a cut-vertex? a bridge?

7. Define a *perfect graph* as a graph all of whose degrees are distinct. Prove there is no nontrivial perfect graph.

8. Design algorithms for converting from one graphical representation to another, for adjacency matrix, edge list, and adjacency list.

9. Design an algorithm for computing $G - v$, $G - (u, v)$, $G \cup \{v\}$, and $G \cup (u, v)$ with respect to each representation.

10. Write an algorithm that generates random graphs of given edge density, that is, with a given value of $2|E|/(|V|(|V| - 1))$.

11. Write an algorithm that generates random connected r-regular graphs.

12. Derive the $O(|V|^4)$ estimate for the time required to calculate the reachability matrix.

13. Show there are only five regular planar graphs in which each face has the same number of bounding edges.

14. What can you say about a cubic graph that satisfies $|E| = 2|V| - 3$?

15. Can a vertex be a cut-vertex in both G and G^c?

16. Let $G(V, E)$ be a graph with weights on its edges. Prove that if M_1 has maximum edge weight among all matchings with k edges, and if p is an augmenting path with respect to M_1 of maximum edge weight, the matching M_2 obtained by augmenting M_1 using p has maximum edge weight among all matchings with $k + 1$ edges.

17. Prove the given alternating path algorithm does not work if the graph has odd cycles by giving a counterexample.

18. Prove that if M_1 and M_2 are edge disjoint matchings in a graph $G(V, E)$, there are matchings M_1' and M_2' such that $|\mathbf{M}_1'| = |M_1| - 1$, $|M_2'| = |M_2| + 1$, and $M_1 \cup M_2 = M_1' \cup M_2'$.

19. Determine the performance of Demoucron's algorithm, assuming some suitable data structures and using the known bounds on the size of planar graphs.

20. Use Demoucron's algorithm to establish the nonplanarity of $K(3, 3)$ and $K(5)$.

21. Prove that for $|V| \geq 9$, G or G^c is non-planar; while for $|V| < 8$, G or G^c is planar. The problem is difficult when $|V| = 9$ or 10.

22. Can you design an $O(|V|^6)$ algorithm to determine planarity using Kuratowski's theorem?

23. Embed $K(5)$ on the Mobius strip.

24. Construct a planar graph G with $min(G) \geq 5$.

25. Show that every planar graph of order at least 4 has at least four vertices of degree at most 5.

26. Prove there is no planar map with five regions in which every pair of regions is adjacent.

27. If $G(V, E)$ is planar and girth(G) equals $k \geq 3$, then $|E| \leq k(|V| - 2)/(k - 2)$. Use this result to show that Petersen's graph is nonplanar.

28. Let $G(V, E)$ be a nonseparable planar graph. Prove that G is bipartite if and only if the dual of G is eulerian.

29. Represent $K(10)$ as the union of three planar graphs.

30. Carefully check that the eulerian trail algorithm is correct. Design an alternative recursive implementation of the algorithm based directly on the inductive proof of the Eulerian Characterization Theorem.

31. Prove the Hamiltonian Closure Theorem.

32. Prove the remaining part of Ore's Theorem.

33. Modify the backtracking algorithm for hamiltonian paths so that it finds a hamiltonian path starting at a given vertex i and fails if there is none. The algorithm should return the next hamiltonian path, with respect to some ordering of the paths, each time it is called.

34. Adapt the hamiltonian path algorithm so it allows one vertex to be covered twice.

35. Prove $K(5, 3)$ has neither a hamiltonian cycle nor path.

36. Show Petersen's graph $P(V, E)$ is not hamiltonian, but $P - v$ is hamiltonian for every vertex v in $V(P)$.

37. Suppose V is a set of people with $|V| \geq 4$. Prove that if for every subset X of 4 people in V, there is someone in X who knows everyone in X, someone in V knows everyone in V.

38. Implement the following procedure of Tutte (1963) for drawing any planar three-connected graph (with straight lines!). Let C be a cycle which bounds a region in some planar drawing of G. Place the vertices of C in order at the vertices of a regular polygon. Then, place each of the other vertices so that it is at the centroid of its adjacent vertices. Finally, connect adjacent vertices by straight lines.

2

Algorithmic Techniques

2-1 DIVIDE AND CONQUER AND PARTITIONING

The technique of solving a problem by breaking it into smaller, more easily solvable parts, solving these partial problems, and then combining the partial solutions into a solution to the whole problem is called *divide and conquer*. This technique is a basic strategy in software engineering where large problems are commonly attacked by being first partitioned into smaller, independent pieces. Applied to the design of algorithms, the method involves partitioning a problem into smaller, homogeneous pieces, to which the overall technique is then applied recursively. If the parts can be identified and the partial solutions combined efficiently, the method may lead to an efficient algorithm.

Partitioning is a related technique, in which a problem is divided into pieces, solutions are obtained for each piece, and are then combined to form a solution for the whole problem. The difference between divide and conquer and partitioning is that divide and conquer typically involves recursion, while partitioning does not.

A sophisticated example of a graph-theoretic divide and conquer algorithm is the planar shortest path algorithm described in Lipton and Tarjan (1977). The algorithm is based on partitioning a planar graph into planar subgraphs of roughly equal size, and then applying the algorithm recursively. This section illustrates the use of partitioning to improve the efficiency of shortest path algorithms on graphs that can be easily disconnected.

Shortest paths by partitioning. Consider the problem of computing all intervertex shortest distances on a digraph $G(V, E)$ which has positive weights assigned to each edge. We define a *vertex separator* for G as a set of vertices S in $V(G)$ whose removal disconnects G. We will show how to partition G into (overlapping) parts using a vertex separator, and then use the shortest distance subproblems on the parts to find the shortest distances on G. For simplicity, we restrict ourselves to the case where S disconnects G into two components, though a similar technique is applicable when S separates G into many parts.

We denote the edge weight for an edge x by $w(x)$, the two components of $G - S$ by $G_1(V_1, E_1)$ and $G_2(V_2, E_2)$, and the subgraphs of G induced by $V_i \cup S$ ($i = 1, 2$) by H_i ($i = 1, 2$). The shortest distances on G can then be calculated as follows.

1. Solve the intervertex shortest distance problem restricted to H_1 (that is, for shortest paths lying in H_1) using one of the methods of Chapter 3.

2. Replace the original edge weights on the induced subgraph on S by the weights obtained from step (1) (that is, if s_1 and s_2 are in S, then set $w(s_1, s_2)$ to the distance between s_1 and s_2 given by step (1)). Then, solve the intervertex shortest distance problem restricted to H_2, using the new edge weights for edges in S. This gives the true distances between vertices in H_2.

3. Replace the revised edge weights on the induced subgraph on S by the weights obtained from step (2) (that is, if s_1 and s_2 are in S, then set $w(s_1, s_2)$ to the distance between s_1 and s_2 given by (2)). Then, solve the intervertex shortest distance problem restricted to H_1, using the revised edge weights for edges in S. This gives the true distances between vertices in H_1.

4. Find the shortest distance from a vertex u in H_1 to a vertex v in H_2 by minimizing $w(u, s) + w(s, v)$ over all vertices s in S, and similarly for the shortest distances from vertices v in H_2 to u in H_1. This gives the true distances from H_i to H_j, $i, j = 1, 2$.

The correctness and performance of this procedure is summarized in the following theorem.

Theorem (Partitioned Shortest Paths).   Let $G(V, E)$ be a weighted digraph with positive edge weights and let S be a vertex separator whose removal separates G into two components $G_1(V_1, E_1)$ and $G_2(V_2, E_2)$. Let H_i be the induced subgraph on $V_i \cup S$ ($i = 1, 2$). Then, the partitioning algorithm described calculates all the correct shortest distances on G in time $O(|V(H_1)|^3 + |V(H_2)|^3 + |V(H_1)|\,|V(H_2)|\,|S|)$.

We leave the proof as an exercise. The cubic terms in the performance bound arise from the estimates for the performance of Floyd's shortest path algorithm given in Chapter 3.

2-2 DYNAMIC PROGRAMMING

Many combinatorial problems have the property that an optimal solution to the problem can be obtained by combining optimal solutions to subproblems. The problem of shortest paths is an example. If $G(V, E)$ is a graph with positive distances assigned to each edge, it is easy to show that the subpaths of a shortest path between a pair of vertices s and t in G are themselves shortest paths between their endpoints. This property can be used to design an algorithm that inductively constructs the shortest path from s to t.

If an optimal solution to a problem can be decomposed into optimal solutions for subproblems, we say the *principle of optimality* holds. This is the defining characteristic of problems solvable by *dynamic programming*. Not every problem has this property. For example, consider the problem of finding shortest paths on graphs that allow negative edge weights. Referring to the graph in Figure 2-1, observe that a-b-c is a

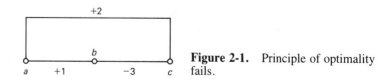

Figure 2-1.   Principle of optimality fails.

shortest path between a and c of length -2. However, unlike for the positive weight shortest path problem, the subpaths of this shortest path are not necessarily shortest paths. Thus, the shortest path between a and b is a-c-b, of length -1, not a-b, of length $+1$. The principle of optimality fails to hold for this problem, since the component solutions of an optimal solution are not necessarily optimal subsolutions. For the principle of optimality to hold, a problem must in some sense be separable into subproblems which can be solved independently, and which have the property that an optimal solution to the original problem is a composition of optimal solutions to a set of subproblems.

Knapsack problem.   As another example, consider the so-called *Knapsack Problem,* wherein we seek to maximize $v_1 x_1 + \ldots + v_n x_n$, where the v_i's are positive real numbers, and the x_i's are nonnegative integers subject to the constraint that $c_1 x_1 + \ldots + c_n x_n \le b$, where the c_i's are also positive reals and b is a fixed positive integer. The name of the problem derives from the following interpretation. A knapsack is to be packed with the most valuable combination of a set of objects, but subject to a constraint on the total weight of the combination. A single instance of object i has value v_i and weight c_i, and b denotes the limit on the total weight. The integers x_i give the number of objects of type i to be selected. The problem obviously satisfies the principle of optimality. For example, if we remove any object i from a valid optimal knapsack, the objects that remain must form an optimal knapsack with respect to the weight limit $b - c_i$. Otherwise, we could obtain a better solution by improving the configuration of objects in the smaller sack. We can use this observation to design a dynamic programming algorithm for the problem as follows. Let $f(k, a) = \max\{v_1 x_1 + \ldots + v_k x_k\}$ subject to the constraint that $c_1 x_1 + \ldots + c_k x_k \le a$, where $0 \le k \le n$ and $0 \le a \le b$. That is, $f(k, a)$ is the maximum value obtainable using the first k objects and a weight limit of a; f satisfies the following recursion

$$f(k, a) = \max\{f(k - 1, a), f(k, a - c_k) + v_k\}$$

by a simple application of the principle of optimality. The objective function whose value is to be optimized is just $f(n, b)$. This quantity depends recursively on at most nb previously computed terms: $f(0, a), 0 \le a \le b, f(1, a), 0 \le a \le b, \ldots, f(k, a), 0 \le a \le b - c_k$. It is a characteristic feature of dynamic programming that, rather than implementing the algorithm recursively, as suggested by the recurrence relation, we implement it in an iterative manner, computing and retaining the values of the terms $f(i, a)$ as they are computed. This eliminates the need for a wasteful recomputation of these values and is the basis of the efficiency of the method.

Multi-stage shortest paths.   A *multi-stage digraph* is a k-partite directed graph $G(V_1, \ldots, V_k)$ where every edge is between a part V_i and its successor part V_{i+1}. We will assume that $V_1 = \{s\}$ and $V_k = \{t\}$, and that each edge (i, j) has a real-valued weight denoted by $w(i, j)$. The length of a path is defined as the sum of the weights of the edges on the path. The shortest path from a vertex u to a vertex v is the path from u to v of minimum length $W(u, v)$.

A shortest path from s to t can be computed using dynamic programming, since the shortest paths satisfy the following optimality principle: a shortest path from s to a vertex u in part V_{i+1} consists of a shortest path from s to some vertex v in part V_i followed by the edge (v, u). Thus, if x is in V_{i+1}, then $W(s, x)$ satisfies the recurrence relation

$$W(s, x) = \min_{y \in V_i} \{W(s, y) + w(y, x)\}.$$

For purposes of efficiency, it is important that we not interpret this simply as a recursion in which $W(s, x)$ is defined in terms of other recursively defined terms $W(s, y)$. It is important that the terms $W(s, y)$ be computed only once. Otherwise, the algorithm may repeatedly and unnecessarily recompute terms such as $W(s, y)$, such as for the other pairs (s, x') where x' is in V_{i+1} and y is adjacent to x'. Such computations in turn would trigger further recomputations of similar terms all the way down through the lower levels of the graph. The expression sought is

$$W(s, t) = \min_{y \in V_{k-1}} \{W(s, y) + w(y, t)\}.$$

The ith stage requires as many as $|V_i||V_{i-1}|$ computations which are based on $|V_{i-1}|$ previously computed and stored terms. If we sum this over every stage, we find that $W(s, t)$ requires

$$\sum_{i=2}^{k} |V_i||V_{i-1}|$$

computations.

We can generalize the problem by allowing the graph to contain edges from any stage V_i to any later stage V_j, where $j > i$. Then, if x lies in part V_i, $W(s, x)$ satisfies

$$W(s, x) = \min_{y \in V_j, \, j<i} \{W(s, y) + w(y, x)\}.$$

The cost of the shortest path algorithm is correspondingly greater. The time to compute the distances for V_i is as great as

$$\sum_{j<i} |V_i||V_j|.$$

The requirement to retain previously computed terms is even more stringent. Thus, altogether,

$$\sum_{i=1}^{k-1} |V_i|$$

terms must be retained, which may be $O(|V(G)|)$.

Optimal program segmentation. We now consider the problem of partitioning a program in a paged memory environment in a way that minimizes the number of interpage transfers of control generated during the execution of the program. The problem can be solved using dynamic programming on a multistage, weighted digraph model (see Kernighan [1971]).

We model the program as a digraph $G(V, E)$ defined in terms of the blocks of the program. A *block (of code)* is a contiguous sequence of code in a program which is executed as a unit. A block has the property that either every statement in the block is executed or none is, and so it represents a natural point of reference for modeling transfer of control. We let the vertices of G correspond to the blocks of the program, while

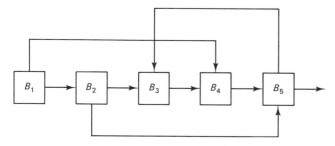

(a) Code block structure of program.

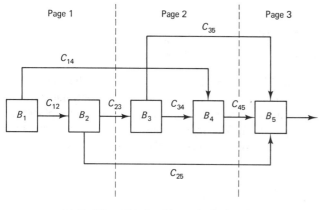

(b) Partition of blocks with transfer frequencies.

Figure 2-2. Optimal partitioning problem.

the edges of G correspond to the possible transfers of control from one block of code to another. Assuming a statistical profile of the branching behavior of the program is available, we can assign a weight to each edge of G which equals the expected frequency of execution of the branch represented by the edge. Branches can be in either direction between a pair of blocks, and so we will combine the frequencies in both directions and assign it to a single edge in the forward direction (with respect to the natural order of the blocks in the program). The vertices of G are assigned weights equal to the lengths of the blocks of code they represent. Refer to Figure 2-2 for an example.

We denote the blocks, listed according to their program order, by B_i, $i = 0, \ldots,$ $N + 1$, where for convenience in the formulation of the dynamic programming algorithm we have introduced two artificial blocks, B_0 and B_{N+1}, of zero weight, which have no transfers into or out of them. The *size of block* B_i is denoted by W_i. The *weight of an edge* from B_i to B_j is denoted by w_{ij}. Each *page* has the same fixed *size* (or *weight*) p. An integral number of blocks can be packed per page, subject to the constraint that their total weight does not exceed p. Blocks are stored sequentially within a page, that is, all the blocks in a page correspond to an ordered sequence $B_i, \ldots, B_{i+k}$. The objective is equivalent to allocating the blocks to pages, consistent with the assumptions, to minimize the total weight of the edges between the pages (the *cost* of the partition).

The dynamic programming solution is defined in terms of breakpoints, where a *breakpoint* is simply the index of the first block on a page. A partition is determined by

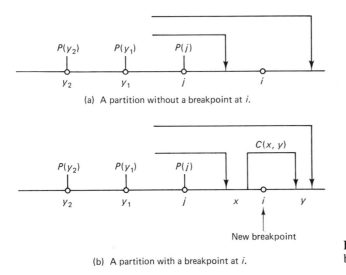

(a) A partition without a breakpoint at i.

$C(x, y)$

New breakpoint

(b) A partition with a breakpoint at i.

Figure 2-3. Partitions defined by breakpoints.

its sequence of breakpoints; refer to Figure 2-3a for an example. The distance $D(i,j)$ from a block B_i to B_j is defined as $W_i + \ldots + W_j$. The cost of an optimal partition whose last breakpoint is i is denoted by cost(i), no weight constraint being placed on the final segment $[i, N+1]$ of the partition. The incremental (!) cost of introducing a further (final) breakpoint at i, given that the previous (final) breakpoint was at j, is denoted by $C(i,j)$. Algebraically,

$$C(i,j) = \sum_{\substack{j \leq x < i \\ y \geq i}} w_{xy} .$$

Refer to Figure 2-3 for a geometric illustration of the meaning of $C(i,j)$.

We now give a dynamic programming algorithm for an optimal partition which returns the cost of an optimal partition in cost($N + 1$). The optimal partition itself can be determined from the terms pred(i) in the algorithm by tracing the breakpoint values back from pred($N + 1$), as illustrated by the example in Figure 2-4. The procedure is

(1) Set cost(1) to 0.
(2) for $i = 2$ to $N + 1$ **do**
　　Set cost(i) to min $\{$cost(j) + C(i,j) $|$ D($j, i-1$) $\leq p\}$
　　Set pred(i) to the smallest value of j that minimizes the above expression.

The correctness of the procedure is established by the following theorem.

Theorem (Correctness of Optimal Partition). Let $G(V, E)$ be a weighted graph which models the blocks of a program. Then, the partitioning algorithm correctly computes both the cost and the parts of an optimal partition on G.

To prove the correctness of the algorithm, we argue as follows. Let Par(i) denote the set of breakpoints of an arbitrary partition whose last breakpoint equals i. Let Opt(i) denote the set of breakpoints of an optimal partition whose last breakpoint equals i and whose cost is cost(i). Opt(i) must equal Par(j) $\cup \{i\}$, for some breakpoint j less than i, where j lies within distance p of i. The cost of Par(j) $\cup \{i\}$ equals cost(Par(j)) + C(i,j)

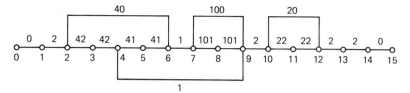

(a) Blocks with their branching frequencies.

cost	pred
cost(1) = 0	pred(1) = 1
cost(2) = 2	pred(2) = 1
cost(3) = 82	pred(3) = 1
cost(4) = 82	pred(4) = 1
cost(5) = 82	pred(5) = 1
cost(6) = 84	pred(6) = 2
cost(7) = 83	pred(7) = 5
cost(8) = 283	pred(8) = 5
cost(9) = 283	pred(9) = 5
cost(10) = 85	pred(10) = 7
cost(11) = 125	pred(11) = 7
cost(12) = 127	pred(12) = 10
cost(13) = 87	pred(13) = 10
cost(14) = 87	pred(14) = 10
cost(15) = 87	pred(15) = 13

(b) Table for cost(x) and pred(x).

Breakpoints: 13, 10, 7, 5

Partition: {1, 2, 3, 4}, {5, 6}, {7, 8, 9}, {10, 11, 12}, {13, 14}

(c) Optimal partition.

Figure 2-4. Optimal partitioning with $p = 4$ and unit block weights.
Source: (a) Taken from Kernighan (1971), copyright ACM Inc.

by the definition of $C(i,j)$. Since the contribution $C(i,j)$ depends only on the break-point i and the predecessor breakpoint j, and is independent of the internal structure of Par(j), we can assume that an optimal partition Opt(i) has the form Opt(j) $\cup$ {i}, for some index j. Otherwise, we could lower the cost of the partition by merely selecting an optimal choice for Par(j). Thus, the principle of optimality applies. Of course, the optimal choice for j is simply that index which minimizes the sum: cost(Opt(j)) + $C(i,j)$, or, equivalently, cost(j) + $C(i,j)$, over the allowed range of j's. If j satisfies these constraints, the resulting partition must be optimal. This completes the proof.

2-3 TREE BASED ALGORITHMS

Many combinatorial algorithms use *tree-like search procedures* to find optimal solutions to problems or to improve the value of suboptimal solutions. A number of such algorithms are described in subsequent chapters. For example, the shortest path algorithm of Dijkstra (Chapter 3) finds the shortest path from an initial vertex s to a target vertex t by extending a shortest path search tree rooted at s until it reaches t, at which point the path through the tree from s to t is a shortest path. The Ford and Fulkerson maximum flow algorithm (Chapter 6) uses a flow augmenting search tree to find a flow augmenting path which can be used to augment the value of a flow on a flow network. We have already seen that alternating search trees with respect to a matching (Chapter 1) can be used to find augmenting paths which can be used to increase the size of a matching. The simpler problem of determining the components of a graph can similarly be solved by repeatedly extending a search tree from an initial vertex s until the search tree spans the component s lies in. A parallel version of such an algorithm is given in Chapter 9.

The general format of a tree search procedure is as follows. It begins at an initial vertex, and advances until it comes to some special vertex. Because the search usually lacks a global criterion to guide its direction, its advance is based on purely local information, such as, whether an edge or a vertex has been visited previously or is useful to the optimization. The search never advances to a vertex that has already been explored in order to avoid reexploring areas that have already been traversed. This gives the search its tree-like character, since the edges along which the search advances, as opposed to those which are examined but not followed, induce a tree in the graph being processed. As the search progresses, it eventually becomes blocked and is forced to alter its direction. If the blockage is in a particular direction only, the algorithm can try to advance the search from some previously reached point. On the other hand, a complete blockage usually signals termination of at least the current phase of the algorithm. Thus, a search depends on the criteria used to determine whether to advance the search tree to a given vertex or along an edge, and such other determinants as how are incompletely processed vertices or bypassed edges backlogged or rescheduled for subsequent traversal and, how are breakthroughs to a special vertex, and partial or complete blockages handled?

The criteria that control a search are problem dependent. For example, the decision to advance along an edge may depend on whether the edge has a capacity for additional flow or an alternating character with respect to a matching. The techniques for scheduling backlogged vertices for further processing are also problem-dependent, and include queues (for breadth first search (Chapter 5)), or stacks (for depth first search (Chapter 5)), or some other discipline. Critical actions often occur when the search is redirected. For example, if parameters have been assigned to vertices as the search advanced, on redirection they may be tested against some criterion which triggers an action when successful (such as the listing of blocks, or strong components in Chapter 5). A breakthrough to a special vertex usually indicates the detection of a path from the root of the tree through the tree to the special vertex, which may be the solution to the problem, or an aid in improving a current suboptimal solution. On the other hand, a complete blockage of the search from its current starting point, though in general indicating termination of at least the current phase of the algorithm, also usually has some problem-dependent implication, such as that the subgraph reached by the search is irrelevant to the optimization problem (such as an Hungarian tree for the matching problems of Chapters 1 and 8), or that there is no solution to the problem (such as a failure to reach the destination vertex in Dijkstra's shortest path algorithm). The blocked search tree itself may have an important problem interpretation (such as defining a minimum cut in the Ford and Fulkerson flow algorithm in Chapter 6).

The most prominent graph search technique is depth first search, which can be used to identify a variety of structural properties of a graph, such as its orientability, the identity of its strong components or blocks, or to test for the existence of cycles (Chapter 5). Other applications of tree-like procedures include data distribution on a parallel computer (Chapter 9), where data is disseminated around a network in a tree-like manner to allow parallel computation on the data. In some applications, an actual tree in the problem graph (as opposed to a search tree used to explore the graph) is useful, such as a spanning tree to determine a basis for the cycles in a graph (Chapter 4). In contrast to tree search, certain problems use other types of search, such as the layered search network approach for Dinic's flow algorithm (Chapter 6), where the objective is not just to reach a special vertex in the most economical way, but to use the intermediate connections to the fullest extent possible. In particular, edges that visit previously reached vertices may still be useful, leading to a non-tree-like search.

2-4 BACKTRACKING

The solution to many combinatorial problems consists in finding an optimal configuration with a desired property in a universe of possible configurations. The brute force implementation of a search for a configuration is called *exhaustive search* and is usually computationally cost-prohibitive, except for very small scale problems. *Backtracking* is a modified version of exhaustive search which takes advantage of some special property of the problem to avoid having to examine every possible configuration. The idea of backtracking is to try to repeatedly extend a partial solution to the problem, represented as a vector, to a complete solution to the problem, by extending the vector representation of the solution one element at a time, whenever possible, and shrinking the solution vector ("backtracking"), whenever a partial solution cannot be extended further.

A backtrack procedure for finding a hamiltonian path in a graph was given in Chapter 1. The format of that procedure is generally typical of backtracking. A backtrack procedure for generating the colorings of a graph is given in Chapter 7. Backtracking can be interpreted as a tree-like search technique, though in a different sense than the techniques of the preceding section. There, a search process explored a problem graph and induced a tree in the graph. In backtracking, the search is through a solution space associated with the problem, and the search tree is a conceptual model of the search process. Chapter 7 discusses this interpretation of backtracking further.

Backtracking models the solution to a problem as a vector $(a_1, \ldots, a_n)$ with elements a_i chosen from finite sets A_i. The search space consists of the cartesian product of the A_i's, $A_1 \times \ldots \times A_n$, which the backtracking algorithm examines in a lexicographically ordered manner. The objective is to find one (or all) solution(s) $(a_1, \ldots, a_n)$ satisfying a given property (or *predicate*) $P_n(a_1, \ldots, a_n)$. At each stage, a partial vector $(a_1, \ldots, a_k)$ (called a *feasible solution*) satisfying a restricted version $P_k(a_1, \ldots, a_k)$ of the property is considered for extension to a partial vector $(a_1, \ldots, a_k, a_{k+1})$ satisfying $P_{k+1}(a_1, \ldots, a_k, a_{k+1})$. The candidates a_{k+1} for extension are selected from a subset S_{k+1} of A_{k+1} such that $P_{k+1}(a_1, \ldots, a_k, a_{k+1})$ holds. If S_{k+1} is empty, we are forced to backtrack to the shorter partial solution $(a_1, \ldots, a_{k-1})$, since all the possible extensions of $(a_1, \ldots, a_k)$ are precluded as solutions. The method presumes that the predicates satisfy the *domino property*: that is, if property P_k fails, then property P_{k+1} also fails. That is, if a predicate fails for a given partial vector, then the predicates for all extensions of that vector also fail. In summary, backtracking is like a depth first traversal of an abstract search tree, where certain subtrees are excluded from consideration because their leading partial solutions are recognizably deadends.

The *n-queens* puzzle from chess is a standard example of a problem solvable by backtracking. The problem is to place n chess queens on an n by n chess board in such a way that no one of the queens attacks any other. To set the problem up for backtracking, we observe that no column of the board can contain more than one queen. Therefore, we can model the placement of the queens by a vector $(q_1, \ldots, q_n)$, where q_i gives the index of the row where the queen for the ith column is placed. The predicate $P_k(q_1, \ldots, q_k)$ succeeds if the placement of the first k queens is valid, and fails otherwise. This predicate obviously satisfies the domino property. An extension of $(q_1, \ldots, q_k)$ using q_{k+1} is valid if the queen placed in row q_{k+1} attacks none of the earlier queens. The lexicographically smallest partial vector is (1) (or $(1, 0, \ldots, 0)$ if we pad the vector with 0's). For the case $n = 4$, the vector $(1, 3)$ is the first partial solution backtracked from, while the first complete solution is $(2, 4, 1, 3)$.

Backtracking can be used to solve linear algebraic optimization problems such as, find a vector (x_1, x_2, x_3) satisfying $x_1 + 2x_2 + 3x_3 \leq 10$, where $1 \leq x_i \leq 4$, x_i integer. The predicate P_i corresponds to restricting the first inequality to its first i terms only. In this problem, the predicates satisfy the domino property, so backtracking is applicable. On the other hand, the domino property fails for the following problem, find a vector (x_1, x_2, x_3) satisfying $x_1 + 2x_2 - 3x_3 \leq 10$, where $1 \leq x_i \leq 4$, x_i integer, because while the predicate $P_2(4, 4)$ fails, the extended predicate $P_2(4, 4, 4)$ succeeds. However, the problem can be easily modified so the domino property holds.

The domino property allows us to preclude certain solutions from examination. This phenomenon is called *pruning* because the precluded partial solutions form a subtree of the backtrack tree. *Branch-and-bound* is a generalization of this approach. Thus, suppose the objective is to find a solution of minimum cost to some problem and that we define predicates which measure the cost of the current partial solution against the cost c of the least cost (complete) solution discovered so far. Then, a predicate $P_k(a_1, \ldots, a_k)$ succeeds if a feasible solution $(a_1, \ldots, a_k)$ has cost at most c, and fails otherwise. We assume the following version of the domino principle holds: the cost of a partial vector is less than the cost of any possible extension of that partial vector. The traveling salesman problem (find a minimum cost spanning cycle on a complete graph with positive edge weights) is an example of a problem that can be solved using branch-and-bound. A partial solution for this problem consists of a vector representing a path of weight w. If the least cost solution (that is, the least cost spanning cycle) found so far has weight c, any extension to a partial solution of weight more than c is automatically precluded, because it cannot possibly lead to a solution of cost lower than c.

Monte Carlo estimate of backtracking performance. One measure of the performance of a backtracking algorithm is the number of vertices in the backtrack search tree. Backtrack algorithms are expected to have exponential performance with respect to this measure, since their solution vectors typically have the form $(a_1, \ldots, a_n)$, where a_i lies in A_i, which leads to a cartesian product search space with $|A_1| \times \ldots \times |A_n|$ vertices. Even if pruning and other techniques reduce the average size of A_i to a constant c, there may still be c^n vertices in the tree. However, in practice, backtracking algorithms often have much better behavior.

Monte Carlo estimation is a method of estimating the size of a pruned backtrack search tree. The idea is to pick a random path through a search tree and to suppose that the length and the branching behavior of that path are typical of the tree in general. Thus, suppose there are m_1 choices for a_1. Select a value for a_1 at random and let m_2 equal the resulting number of feasible candidates for a_2, that is, such that (a_1, a_2) is feasible. Randomly select one of the m_2 feasible values for a_2, and let m_3 equal the number of resulting feasible values for a_3, that is, such that (a_1, a_2, a_3) is feasible. Define m_i for $i > 3$ similarly. Let the length of the random path P generated in this manner equal k. Since there are m_i possible branches at each point of P, we estimate the number of vertices in the tree as

$$\prod_{j=1}^{k} m_j .$$

Recursive backtracking algorithm for digraph isomorphism.

A pair of digraphs G_1 and G_2 are said to be *isomorphic* if there is a 1-to-1 mapping between their vertices which preserves adjacency. We will describe a recursive backtracking procedure Isomorphic for digraph isomorphism. The procedure is actually a function Isomorphic(G_1,G_2,Map,S) which is invoked with Isomorphic initialized to false, the set S initialized to empty, a global variable i initialized to 0, and a vector Map in which the isomorphism is returned initialized to 0. A subordinate function Test(Map,i,j) implements the obvious partial isomorphism predicate by merely testing if mapping i in G_1 to j in G_2 satisfies the adjacency constraints requirements for isomorphism. That is, that i in G_1 is adjacent to i' ($<$i) in G_1 if and only if j in G_2 is adjacent to Map(i') in G_2. We assume $|V(G_1)|$ equals $|V(G_2)|$.

```
Function Isomorphic(G₁,G₂,Map,S)

(* Returns isomorphism between G₁ and G₂ in Map, or fails *)

var G₁, G₂: Graph
    i,j: 0..|V|
    S: Set of 1..|V|
    Map (1..|V|): 0..|V|
    Isomorphic, Test: Boolean function

Set i to i + 1
if   S = V then Set Isomorphic to True

for   j in V(G₂) − S
  do   while   not Isomorphic
           do   if Test(Map,i,j)
                  then   set Map(i) to j
                         Isomorphic(G₁,G₂,Map,S ∪ {j})

Set i to i − 1

End_Function_Isomorphic
```

We can improve the performance of the algorithm by testing additional constraints in Test(Map,i,j). For example, we could check whether the in-degree and out-degree of vertices i and j were equal. It might also be useful to order the vertices of G_1 so that vertices near the root of the backtrack tree have smaller degree. For example, we could select for first examination, that vertex x in G_1 with the property that the smallest possible number of vertices in G_2 have the same in-degree and out-degree as x. The motivation for this order of processing is that pruning tends to be done near the top of a backtrack tree, since it tends (heuristically speaking) to be recognized after a few constraints have been accumulated. Since the total number of points in the search space is fixed, pruning a backtrack tree with a low degree near the top will tend to maximize the number of search space points that are pruned.

2-5 RECURSION

Recursion is a generalized form of mathematical induction encompassing dynamic programming, divide and conquer, and backtracking, but not limited to these methods. In a recursive algorithm for a graph $G(V, E)$, a problem for G is converted into problems for a set of related graphs $f_i(G)$ which have the property that there is a function c mapping graphs to cardinal numbers such that $c(G) > c(f_i(G))$, and, furthermore, there are a finite (small) number of graphs that c maps to 0. In other words, a recursive graph algorithm reduces a problem about a graph to a problem about some small number of "smaller" graphs, and the recursion bottoms out with a limited number of small special cases that can be treated in an ad hoc manner.

The classical instance of a recursive graph processing algorithm is depth-first search. As mentioned in Section 2-3, a number of important graph processing algorithms are based on enhancements of depth-first search: the determination of blocks, strong components, etc. Recursive and nonrecursive (explicit stack) versions of these algorithms are given in Chapter 5. Depth-first search has an efficient $O(|E(G)|)$ performance, and many of the algorithms based on it inherit this efficiency. But, not every recursive algorithm is efficient. Indeed, a straightforward application of recursion may lead to a very inefficient algorithm. This section gives some elementary examples of recursion and briefly discusses the solution of the recurrence relations associated with the analysis of the complexity of recursive algorithms. For more substantial recursive algorithms, we refer to the various depth-first search based algorithms in Chapter 5.

Vertex connectivity. The appealing simplicity with which a recursive algorithm can be stated may unfortunately be outweighed by drastically poor performance. A recursive algorithm for the vertex connectivity VC of a graph G illustrates this. VC(G) satisfies the simple recurrence relation

$$\text{VC}(G) = 1 + \min_{v \in V(G)} \{\text{VC}(G - v)\}.$$

While similar to a dynamic programming recurrence, this relation makes no use of the principle of optimality, but merely reflects a simple structural relation that VC(G) satisfies. It leads to an obvious but inefficient recursive algorithm for VC(G). If we let $T(|V(G)|)$ denote the time to compute VC(G) using that algorithm, then $T(|V(G)|)$ satisfies

$$T(|V(G)|) = |V(G)|T(|V(G)| - 1),$$

which has an $O(|V(G)|!)$ solution.

The problem with this algorithm is that it is invoked for $|V(G)|$ subproblems of almost the same size as the original problem. If there were a way to efficiently identify a vertex v that attained the minimum value used by the algorithm, such as in linear time $O(|V(G)|)$, eliminating the need to examine the $|V(G)|$ subproblems, then the resulting performance function would satisfy

$$T(|V(G)|) = T(|V(G)| - 1) + c|V(G)|$$

for a constant c, leading to an algorithm with quadratic performance. However, there appears to be no quick way to find the required special vertex, so the recursive approach to this problem seems unappealing. We will describe efficient alternative connectivity algorithms based on maximum flow techniques in Chapter 6.

There is a recursive algorithm for five-coloring a planar graph $G(V, E)$ based on the ability to locate a critical vertex v of degree at most 5 in G in linear time, recursively color $G - v$, and then efficiently modify that coloring to a five-coloring of G. The associated recurrence relation has the additive linear form that leads to an algorithm quadratic in $|V(G)|$. The algorithm is described in Chapter 7.

Chromatic polynomial. Another, rather theoretical, example of a problem with a recursive solution, is the problem of computing the chromatic polynomial of a graph G. This algorithm also has exponential performance, but this is unremarkable since the problem itself appears to be intrinsically exponential. We define an *x-coloring* of a graph G to be an assignment of at most x different colors (or integers on $[1, x]$) to the vertices of G so adjacent vertices have different colors. A pair of x-colorings are considered different if they assign different colors to some vertex. This terminology is discussed in greater detail in Chapter 7. We denote the number of different x-colorings on G by $\mathrm{chr}(G, x)$. If x is less than the vertex chromatic number of G, then $\mathrm{chr}(G, x)$ is zero. However, in general, $\mathrm{chr}(G, x)$ is a polynomial of degree $|V(G)|$ in x, called the *chromatic polynomial* of G.

A backtracking algorithm for finding every coloring of a graph is given in Chapter 7 which could be used to calculate the chromatic polynomial as a by-product. Conversely, the chromatic polynomial could be used to estimate the size of a backtrack tree, since it equals the number of successful (complete) search paths in the tree, and so gives a lower bound on the number of vertices in the tree. It is at least as costly to compute the chromatic polynomial as to solve the minimization problem: find the minimum number of colors needed to color the vertices of a graph. In the terminology of Chapter 10, the latter problem is already NP-Complete, meaning that it is unlikely to have a polynomial time solution. Since we can use the chromatic polynomial $\mathrm{chr}(G, x)$ to solve the minimum coloring problem, it is at least as difficult (that is, probably exponential) to compute the chromatic polynomial as to solve the minimization problem.

The algorithm depends on a relation between the chromatic polynomial and certain reductions of a graph. If $G(V, E)$ is a graph and $x = (u, v)$ is an edge of G, we define G/x to be the graph obtained from G by collapsing x so that u and v merge into a single vertex. The resulting self-loop is deleted, and any parallel edges are replaced by a single edge.

Theorem (Chromatic Polynomial Recursion). Let $G(V, E)$ be a connected graph, and let (u, v) be in $E(G)$. Then, the chromatic polynomial $\mathrm{chr}(G, x)$ satisfies

$$\mathrm{chr}(G, x) = \mathrm{chr}(G - \{(u, v)\}, x) - \mathrm{chr}(G/\{u, v\}, x).$$

The proof of the theorem is trivial. Observe that the difference $\text{chr}(G - \{(u, v)\}, x) - \text{chr}(G, x)$ is exactly the number of x colorings of G which would be valid except that u and v are colored the same. But, these are precisely the valid x colorings of $G/\{(u, v)\}$. This completes the proof.

The theorem expresses $\text{chr}(G, x)$ in terms of the chromatic polynomials of graphs that are "smaller" than G: $G - \{(u, v)\}$ has one less edge than G and no more vertices, and $G/\{(u, v)\}$ has one less vertex and at least one less edge. We can implement the recursion implied by the theorem by removing edges in such a way that we eventually end up with trees as the bottom level graphs of the recursion. Then, we can invoke the following explicit expression for the chromatic polynomial of trees.

Theorem (Chromatic Polynomial of Tree).   Let T be a tree of order n, and let $\text{chr}(T, x)$ be the chromatic polynomial of T. Then, $\text{chr}(T, x)$ satisfies

$$\text{chr}(T, x) = x(x - 1)^{n-1}.$$

Thus, every tree of order n has the same chromatic polynomial.

The proof is by induction. Let u be an endpoint of T and let (u, v) be an edge. $T - \{(u, v)\}$ consists of a tree of order $n - 1$ and an isolated vertex. Therefore, by induction, it has the chromatic polynomial

$$\text{chr}(T - \{(u, v)\}) = x^2(x - 1)^{n-2}.$$

On the other hand, if we coalesce u and v to produce $T/\{(u, v)\}$ we obtain a tree of order $n - 1$, which by induction has chromatic polynomial

$$\text{chr}(T/\{(u, v)\}, x) = x(x - 1)^{n-2}.$$

Applying the Chromatic Polynomial Recursion Theorem, it follows that $\text{chr}(G, x)$ has the desired form. This completes the proof.

Solution of recurrence relations.   Recurrence relations arise naturally in the analysis of the performance of recursive algorithms. If $T(n)$ denotes the time required by an algorithm for an input problem of size n, recursion leads to an expression for $T(n)$ in terms of the performance $T(m)$ of problems of smaller size $m < n$. The recurrence relations come in many forms. The basic method of solution is to iteratively feed the recurrence back into itself until the recursion bottoms out with small cases which can be solved directly.

A common example is for a recurrence of the following form. Suppose that a problem of size n can be solved by being split up (divided) into a problems each of size n/c, which can then be solved recursively by the same method (conquered), and then merged into a solution to the whole problem, where the cost of dividing the problem and merging the solutions to the subproblems is $f(n)$. If we denote the time to solve a problem of size n by $T(n)$, then $T(n)$ satisfies the recurrence

$$T(n) = aT(n/c) + f(n), \qquad n > 1.$$

In the special case that $f(n)$ is linear, the recurrence has the form

$$T(n) = aT(n/c) + bn, \quad \text{for } n > 1$$

where we assume that $T(1)$ equals b. Without serious loss of generality, we will also assume n is a power of c. Then, it can be shown that T has the form

$$T(n) = bn \sum_{j=0}^{\log_c n} (a/c)^j.$$

The behavior of the solution depends on the ratio a/c. If $a/c < 1$, the summation converges and $T(n)$ is $O(n)$. If a equals c, the summation equals $\log_c n$, and $T(n)$ is $O(n \log n)$. If $a > c$, then $T(n)$ equals $O(n^{\log_c a})$, which is superlinear in n. That is,

$$T(n) = O(n), \qquad \text{for } a < c,$$

$$T(n) = O(n \log n), \quad \text{for } a = c,$$

$$T(n) = O(n^{\log_c a}), \qquad \text{for } a > c.$$

A slightly more general recurrence is

$$T(n) = aT(n/c) + O(n^d), \quad \text{for } n > 1.$$

In this case, the solution has the form $O(n^d)$ if $\log_c a < d$; $O(n^d \log n)$ if $\log_c a = d$; and $O(n^{\log_c a})$ if $\log_c a > d$, which is greater than $O(n^d)$. That is,

$$T(n) = O(n^d), \qquad \text{for } a < c^d,$$

$$T(n) = O(n^d \log n), \quad \text{for } a = c^d,$$

$$T(n) = O(n^{\log_c a}), \qquad \text{for } a > c^d.$$

There are similar results when the recurrence relation has some other simple forms and a readily accessible general theory as well. We refer the reader to the references.

Knowledge of the behavior of these recurrences can be used to guide the process of algorithm design. Thus, if $\log_c a > d$, the speed of the algorithm can be increased by shrinking the ratio a/c, that is, making the number of parts and/or the size of the parts smaller. On the other hand, if $\log_c a < d$, then to increase the speed of the algorithm, we should try to decrease the cost of the overhead of dividing and merging.

2-6 GREEDY ALGORITHMS

The *greedy method* is the most primitive algorithmic design technique. The idea is to generate a solution in a step by step manner by making the maximum possible improvement in an objective function at each step. Thus, if a problem requires a solution S optimizing the value of an objective function f under a constraint C, a greedy algorithm for the problem constructs a sequence of subsolutions S_i, $i = 1, \ldots, n$, each subsolution satisfying a possibly restricted version of the constraint, and where solution S_{i+1} is obtained by extending solution S_i in a way that maximizes the possible one-step improvement in f (or, equivalently, minimizes the increase in f in the case that f is to be minimized), and such that S_n equals S.

The traveling salesman problem, which seeks a minimum weight spanning cycle in a weighted graph, has a simple greedy algorithm: select an initial vertex v_0; travel to the nearest unvisited neighbor v_1 of v_0; then, travel to the nearest unvisited neighbor of v_1, and so on, taking care not to return to the initial vertex until after every other vertex has been visited. The constraint is to keep the path (extendible to) a spanning path.

Subject to this constraint, the objective function (the path length) is minimally increased at each step. The algorithm is fast, but it is easy to construct examples that show it is not always optimal. Indeed, the solution can be arbitrarily worse than optimal in the sense of relative error. Despite its limitations, this kind of local optimizing technique is frequently useful. The minimum spanning tree problem of Chapter 4 is an example of a problem which has a simple greedy algorithm which yields an optimal solution. We will illustrate the greedy method here with an exact optimal record layout algorithm, and an approximately optimal algorithm for a search tree problem.

Optimal record layout. Consider the problem of how to optimally arrange records on a linear storage device, where the frequency with which each record will be accessed is known beforehand. The objective is to arrange the records in a way that minimizes the average distance traveled by the access head as it moves from one record to the next requested record. We model the storage medium as a real line with the records positioned at points with integer coordinates, at most one record per point. The access head remains at the position of the last record requested. Access requests are placed on an I/O request queue as they arrive, and serviced one by one in a first-in-first-out fashion. The requests are assumed to be statistically independent, and the time to access a record is proportional to the distance of the requested record from the current position of the access head.

Denote the set R of records by $\{r_1, \ldots, r_n\}$ and denote the probability that r_i is requested by p_i. The distance between records r_i and r_j is denoted by $\mathrm{Dist}(i,j)$ and depends on the arrangement A of the records. The metric is the ordinary Euclidean metric. The objective function to be minimized is the expected distance traveled by the access head as it moves from one requested record to the next, that is

$$D(A) = \sum_{i,j} p_i p_j \, \mathrm{Dist}(i,j) \, .$$

A greedy procedure that finds an arrangement of records that minimizes D is as follows: place the most frequently accessed record at 0, the next most frequently accessed at 1, the third most frequent at -1, and so on, alternating the records to the left and right of 0, in order of decreasing frequency. The algorithm is greedy because each successive record placement positions the next most frequently accessed record in a way that minimizes the increase in D. The resulting arrangement of the probability distribution exhibits a pipe-organ pattern. The following theorem establishes its optimality.

Theorem (Optimality of Pipe-organ Arrangement). Let $R = \{r_i, \ i = 1, \ldots, n\}$ be a set of records, where r_i has probability of access p_i. Then, the pipe-organ arrangement generated by the greedy algorithm minimizes the expected distance traveled by the access head between successive record accesses.

First, we shall establish the following lemma.

Lemma. Let A be an optimal arrangement of probabilities and let L be a vertical line through any point of the form $y/2$, for y integral. Then, the probabilities to one side of L are larger than their symmetric probabilities on the other side of A.

The proof is as follows. Assume, to the contrary, that there is a line L such that record q_3, q_1, q_2, q_4 lie in that increasing order, q_1 and q_3 are to the left of L, q_2 and q_4 lie to the right of L, $p(q_1)$ is greater than $p(q_2)$, and $p(q_3)$ is less than or equal to $p(q_4)$. Consider the effect of interchanging q_1 and q_2. The interchange has no effect on those terms in $D(A)$ involving either only q_1 and q_2, or neither q_1 nor q_2. Otherwise, the changes are

Records	Before	After
q_3, q_1	$p(q_3)p(q_1)\mathrm{Dist}(x_3, x_1)$	$p(q_3)p(q_1)\mathrm{Dist}(x_3, x_2)$
q_3, q_2	$p(q_3)p(q_2)\mathrm{Dist}(x_3, x_2)$	$p(q_3)p(q_2)\mathrm{Dist}(x_3, x_1)$
q_4, q_1	$p(q_4)p(q_1)\mathrm{Dist}(x_4, x_1)$	$p(q_4)p(q_1)\mathrm{Dist}(x_4, x_2)$
q_4, q_2	$p(q_4)p(q_2)\mathrm{Dist}(x_4, x_2)$	$p(q_4)p(q_2)\mathrm{Dist}(x_4, x_1)$

Observe that $\mathrm{Dist}(x_3, x_1)$ equals $\mathrm{Dist}(x_4, x_2)$, and $\mathrm{Dist}(x_3, x_2)$ equals $\mathrm{Dist}(x_4, x_1)$. The difference between the sums of the contributions before and after the change equals

$$(p(q_1) - p(q_2))(p(q_4) - p(q_3))(\mathrm{Dist}(x_3, x_2) - \mathrm{Dist}(x_3, x_1)),$$

which is positive since each factor is positive. It follows that interchanging q_1 and q_2 will decrease D.

To complete the proof of the theorem, we observe that an optimal arrangement must have the pipe-organ form, or we can find a line L such that the lemma is violated. We can illustrate the idea by an example. Thus, suppose that r_1 is located at 0, while r_2 is located at 2. Then, $p(r_2)$ is greater than the probability of the record at 1, while $p(r_1)$ is greater than the probability of the record at 3, contrary to the lemma. If r_1 is at 0 and r_2 is at 1, then by the same argument, r_3 must be at the point -1. We can continue this argument to prove the optimal arrangement has a pipe-organ form. This completes the proof of the theorem.

Nearly optimal search trees.   The problem of optimal storage layout is basically free of constraints: any record can be stored in any free location independently of the other records. We now consider the problem of optimally positioning the records in a binary search tree where the record keys are obliged to obey the order constraints of the search tree, and the records are to be arranged in a way that minimizes the average number of comparisons it takes to locate a record. If the access frequencies of the keys are unknown, we merely organize the tree in a balanced fashion. Thus, if there are $2^n - 1$ keys, the records determine a completely balanced binary tree of height n. (Refer to Chapter 4 for this terminology.) If the access frequencies are known, the average access time is minimized by organizing the tree in a manner that depends both on the access probabilities of the keys and the order constraints of the tree.

There are several solutions to this problem. An optimal algorithm of Knuth based on dynamic programming has $O(n^2)$ time. There are two natural greedy algorithms.

Greedy I

(1) Place the most frequent key at the root of the tree, and

(2) Repeat step (1) recursively for the left and right subtrees of the root.

Alternatively, we can keep the tree as frequency balanced as possible.

Greedy II

(1) Select as the root that key whose left and right subtrees are as nearly equal in total frequency of access as possible, and

(2) Repeat step (1) recursively.

It only makes sense to consider the greedy algorithms if they are faster than the optimal algorithm and yield good approximations. The Greedy I algorithm can perform poorly as the following example shows. Let the sorted keys be $k_1, \ldots, k_n$, where $n = 2^k - 1$, with access probabilities $p_i = 2^{-k} + \text{err}_i$, where $\{\text{err}_i, i = 1, \ldots, n\}$ is positive and decreasing and sums to 2^{-k}. Greedy I produces a degenerate tree with k_1 as root and k_i as the right child of k_{i-1} $(i > 1)$. Its average search length is at least $n/2$. On the other hand, the balanced tree with the same probabilities has average search length at most $2\log n$, so that the Greedy I solution is arbitrarily worse than optimal. Greedy II, on the other hand, is both fast (it runs in $O(n \log n)$ time) and has good error behavior. The algorithm is suggested by the optimal configuration for the equal frequency case, where the optimal tree is completely balanced. An analog of this when the frequencies are unequal is the weight balanced tree produced by Greedy II.

We may view the problem in terms of the constraint-local optimization formulation we have described earlier, where the constraint is the search tree order: keys in a left (right) subtree have to be less than (greater than) their root key. The objective function is the average access time for the tree. If there are n keys k_i of access probability p_i, this is defined as

$$\sum_{i=1}^{n} d_i p_i \qquad (2\text{-}1)$$

where d_i equals 1 + the level number of key k_i. An optimal solution configures the tree to minimize (2-1). But, (2-1) depends on the parameters d_i which in turn depend intimately on the structure of the tree, so the expression in (2-1) is not readily calculable. Therefore, we shall work with a modified objective function at each step. Instead of optimizing the improvement in (2-1) at each step, we select a key as root which best balances the weight of the left and right subtrees. Define the weight W of a set of keys $S = \{k_i, i = 1, \ldots, n\}$ as

$$\sum_{i=1}^{n} p(k_i) . \qquad (2\text{-}2)$$

This is just the probability one of the keys in S is accessed. $W(S)$ is 1 if S is the complete set of keys. If a key k_i is selected as the root of the set S, we use as our objective function $f(k_i)$, the difference between the weight $W(S_R(k_i))$ of the set of keys $S_R(k_i)$ greater than k_i, and the weight $W(S_L(k_i))$ of the set of keys $S_L(k_i)$ less than k_i. That key k_i is selected as root which minimizes $f(k_i)$. This determination can be made easily at each step.

Greedy II need not be optimal, because its locally based decisions can constrain subsequent decisions in an unforeseen and suboptimal manner, but empirical studies have demonstrated the goodness of its approximation. Melhorn (1975) first established analytical error bounds. Let H be the entropy of the distribution $\{p_i\}$ defined as

$$\sum_{i=1}^{n} p_i \log(1/p_i) .$$

The Greedy II solution satisfies the following error bound.

Theorem (Melhorn Error Bounds for Greedy II). Let $\{k_i, i = 1, \ldots, n\}$ be a set of keys, each with probability p_i of being accessed. Define $W(\{k_i\})$, as in (2-2), and construct a binary search tree using Greedy II with W as the weight function balanced by the algorithm. Let A_{opt} denote the average access time for an optimal tree, and let A_{bal} denote the average access time for the tree given by Greedy II. Then,

$$.63H \leq A_{opt} \leq A_{bal} \leq 2 + 1.44H .$$

The proof of the theorem is complicated, and we refer the reader to Melhorn's article. The theorem indicates that A_{bal} is never more than a little more than twice as much as A_{opt}.

The time performance of Greedy II is given by the following theorem.

Theorem (Time Performance of Greedy II). Let $\{k_i, i = 1, \ldots, n\}$ be a set of keys, each with probability p_i of being accessed. Then, the Greedy II algorithm can be implemented in $O(n \log n)$ time.

The proof is as follows. We may assume the keys are sorted. Then, we merely find i so the difference between the partial sums

$$\sum_{j=1}^{i-1} p_j - \sum_{j=i+1}^{n} p_j$$

is minimized. This can be done in time $O(\min\{i, n - i + 1\})$ by searching for i alternately from both sides of the sequence. Once i is found, we apply the procedure recursively to the resulting subtrees, which are subproblems of size $i - 1$ and $n - i$. The overall algorithm has performance $T(n)$ which satisfies

$$T(n) \leq \max\{T(i - 1) + T(n - i) + c \min(i, n - i + 1)\}, \qquad 1 \leq i \leq n$$

where c is a constant. We can show $T(n)$ is $O(n \log n)$ as follows.

All the maxima in the following are taken over $0 \leq i \leq n/2$.

$$T(n) \leq \max\{T(i) + T(n - i - 1) + c(i + 1)\},$$

$$\leq \max\{d(1 + i \log i)$$
$$+ d(1 + (n - i - 1) \log(n - i - 1))$$
$$+ d(1 + i)\}$$

$$\leq d(n \log(n) + 1) + d(2 - \log(n - 1))$$
$$+ d \max\{(n - 1)[(i/(n - 1))(\log(i/(n - 1))$$
$$+ ((n - i - 1)/(n - 1))(\log((n - i - 1)/(n - 1))) + (i/(n - 1))]\}$$

$$\leq d(1 + n \log n),$$

where d is a constant, as was to be shown.

2-7 APPROXIMATION

When designing approximate algorithms, we relinquish the quest for an exact solution, in exchange for solutions with good error bounds. We will describe two approximate algorithms for the traveling salesman problem, an algorithm due to Lin which makes iterative local improvements to a feasible solution until a locally optimal solution is obtained and then repeats this procedure for a randomly chosen set of initial solutions, until a good estimate of a globally optimal solution is obtained and an approximation for the euclidean traveling salesman problem based on minimum spanning trees and matchings.

Lin's method. Lin's method combines local optimization, randomization, and experimental tuning, and is supported by extensive experimental work. It generalizes the following simple technique for improving a given spanning cycle.

Let $G(V, E)$ be a weighted graph. The most obvious way to decrease the value of a given spanning cycle C on G is the following criss-cross procedure: given a pair of edges (u_1, u_2) and (v_1, v_2) on C, which lie in the cyclic order u_1, u_2, v_1, v_2, we replace them by the pair (u_1, v_1) and (u_2, v_2) if this reduces the cost of the cycle. The exchange is called an *inversion*. If the new cycle is denoted by C', we can iteratively try to find a pair of edges on C' that can be inverted, to form another spanning cycle C''. The procedure needs to be repeated at most $O(|V|^2)$ times, since if it pays to invert a pair of edges at one point, it cannot later pay to undo the inversion. The entire process is called *2-optimization,* and leads to a spanning cycle which cannot be improved by any inversion, that is, which is *2-optimal.*

We can define *n-optimization* similarly. Experimental results suggest 3-optimization produces better solutions than 2-optimization, while *n*-optimization (for $n > 3$) costs considerably more and yields only marginal improvements.

The procedure for *3-optimization* is defined as follows. We select some combination of three edges from the cycle that have not been selected previously and replace them by three alternative edges, as shown in Figure 2-5. The exchange is called a *3-interchange.* If we denote the selected edges by (u_1, u_2), (v_1, v_2), and (w_1, w_2), we choose the three replacement edges so that the modified cycle is still a spanning cycle, (u_1, w_1), (u_2, v_2), and (v_1, w_2). If this interchange reduces the cost of the cycle, we retain the new cycle; otherwise, we retain the original cycle. The procedure needs to be repeated at most $O(|V|^3)$ times, that is, for every combination of three edges on the successive modified cycles, until a cycle is obtained which is invariant to improvement under *3-interchange.*

In order to obtain a good global estimate of a least cost solution, Lin's method randomly generates initial spanning cycles and 3-optimizes them. The overall least cost

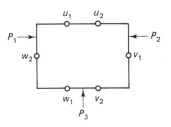

(a) Initial spanning cycle (with constituent paths P_1).

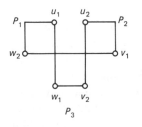

(b) New spanning cycle after 3-interchange.

Figure 2-5. Illustration of three-interchange.

cycle is selected as the optimum cycle. If we visualize the spanning cycles as points x in a multidimensional cycle space, they determine a "surface" $(x, \text{cost}(x))$. The objective is to find the minimum point on this surface. The local optimizing technique is just a hill-climbing method with respect to this surface which identifies local surface optima. Randomizing the choice of the starting point for the search allows us to obtain a cycle which is a good estimate of the global minimum.

Matching method. A previous approximate algorithm for the traveling salesman problem for graphs with euclidean weights depended on converting a minimum spanning tree to an eulerian graph. The relative error was at most 100%. We can improve the bound to a relative error of 50%. The method is as follows. Let $G(V, E)$ be the weighted graph, let T be a minimum spanning tree of G, and let V_0 be the set of vertices in T of odd degree. Since $|V_0|$ is even, we can divide V_0 into two equal parts, V_1 and V_2, of size $k = |V_0|/2$. Let M be a minimum weight perfect matching on the induced complete bipartite graph on V_1 and V_2. There are efficient algorithms for constructing such a minimum weight matching, such as the parallel matching algorithm in Chapter 9. If we add the edges of M to T, allowing parallel edges, we obtain a connected multigraph all of whose vertices have even degree, and which consequently has an edge spanning circuit S. We then use the same technique as in Chapter 1 to reduce S to a spanning cycle. The error bound is given by the following theorem.

Theorem (TSP Error Bound). Let $G(V, E)$ be a Euclidean graph. Let S of weight $\text{cost}(S)$ be a solution to the traveling salesman problem on G. Let A of weight $\text{cost}(A)$ be the spanning cycle constructed by the approximation procedure. Then,

$$\text{cost}(A) \leq (3/2)\,\text{cost}(S).$$

The proof is as follows. Observe that the weight of a minimum weight spanning tree is certainly no more than the cost of an optimal traveling salesman solution, since any spanning cycle contains a spanning tree as a subgraph. Furthermore, the weight of a minimum weight perfect matching on V_0 is at most half the weight of a minimum weight spanning cycle. Thus, let S be an optimal spanning cycle of G. We can convert S to a cycle on the vertices of V_0 by the shortcut procedure for Euclidean weights shown in Section 1-8 and without increasing its weight. By taking every other edge of this cycle, we obtain two perfect matchings on V_0. Since the two matchings together weigh less than the minimum spanning cycle, the shorter of the two matchings cannot weigh more than half the minimum spanning cycle. Therefore, the solution given by the algorithm cannot be more than 50% greater than the weight of an optimal spanning cycle. This completes the proof of the theorem.

2-8 GEOMETRIC METHODS

We usually view graph-theoretic problems from a purely combinatorial perspective, but if the problem has geometric content we may be able to solve it faster using geometric techniques. We will illustrate this approach for a shortest path problem on a planar graph in Chapter 3. Here, we will describe an important general technique for attacking graph-like geometric problems using a geometric construction called the Voronoi diagram. We will show how to construct the Voronoi diagram on N points in the plane in $O(N \log N)$ time and apply it to a simple shortest distance problem. The construction uses an interesting divide and conquer approach. In Chapter 4, we will describe a fast minimum spanning tree algorithm based on the Voronoi diagram.

The *Voronoi Diagram* associated with a set of points S in the plane is defined as follows. Consider for each point s in S, the set of points, denoted by Vor(s, S), which have the property that they are closer to s than to any other point in S. The boundary of Vor(s, S) determines a convex polygon called a *Voronoi polygon*. If we form the Voronoi polygon for each point of S, we determine a partition of the plane that we call the Voronoi diagram, and which we will denote by Vor(S). If there are N points in S, the Voronoi diagram on S contains $O(N)$ edges and N polygons, some of which are bounded and some of which are unbounded.

The Voronoi diagram is algorithmically important because it is a basic geometrical construction and can be computed rapidly. An outline of the construction procedure follows.

(1) Sort the points in S in increasing order of their x coordinates.

(2) Partition S into two parts of size $N/2$, which we denote by L and R, and such that every point in L is to the left of every point in R.

(3) Recursively construct separate Voronoi diagrams Vor(L) and Vor(R) for L and R.

(4) Recombine or merge these separate diagrams to form the desired Voronoi diagram Vor(S).

Step (1) takes $O(N \log N)$ time. Step (2) takes $O(N)$ time. Step (3) is simply a recursion. Step (4) is more complex and is the subject of the following discussion. We will show it can be done in $O(N \log N)$ time.

The key to merging the Voronoi diagrams Vor(L) and Vor(R) in step (4) is the construction of a polygonal line P with the property that all planar points to the left of P are closer to some point of L than to any point of R, while all points lying to the right of P are closer to some point in R than to any point in L. Once we know P, we can merge Vor(L) and Vor(R) by merely superimposing them, eliminating continuations of edges of Vor(L) which lie to the right of P, and continuations of edges of Vor(R) which lie to the left of P.

We will now describe an efficient method for constructing P. Observe that the set of points to the left of P is the union of the Voronoi polygons in Vor(S) whose defining vertices are in L, and similarly for the points to the right of P. Therefore, P is just the set of edges which are shared by the polygons Vor(v, S), for v in L, and Vor(w, S), for w in R. Thus, P consists of a pair of rays (semi-infinite line segments) and a finite sequence of (bounded) line segments. We will construct P by first determining its rays and then use a separate procedure to find its line segments. We use the convex hull of S, denoted CH(S), to construct the rays of P.

An outline of the approach follows. First, we find the rays of P using

(4.1) Vor(L) and Vor(R) to construct CH(L) and CH(R) in linear time, then

(4.2) CH(L) and CH(R) to construct CH$(L \cup R)$ in linear time, and then

(4.3) CH$(L \cup R)$ to construct the rays of P. Once we have the rays of P, we find the remaining line segments of P by

(4.4) Showing how to advance P one line segment at a time.

We will need the convex hull CH(S) for the following reason. A ray of Vor(S) is an unbounded edge bordering an unbounded polygon of Vor(S). Such an edge must be the perpendicular bisector of a pair of consecutive vertices of the convex hull of S. Therefore, we can find the rays of P by finding CH(S).

To construct CH(S), we first show how to construct the convex hull of an arbitrary set in linear time, assuming its Voronoi diagram is given. Since the Voronoi diagram of S is as yet unavailable, we cannot use this method to construct its convex hull directly. But, we can assume that we have already recursively constructed Vor(L) and Vor(R). Therefore, we can construct CH(L) and CH(R) in linear additional time. We then show how to construct the convex hull of the union of two disjoint convex polygons, and therefore CH($L \cup R$), in linear time. Thus, the overall construction of CH(S) takes $O(N \log N)$ time.

Using Vor(L) and Vor(R) to Construct CH(L) and CH(R) [(4.1)].   To construct the convex hull of a set T in linear time $O(|T|)$ from the Voronoi diagram Vor(T), observe that the vertices of CH(T) are precisely the points in T whose Voronoi polygons are unbounded. Since the Voronoi diagram contains only $O(|T|)$ edges, we can identify these polygons by an edge scan that takes only $O(|T|)$ time. It then only remains to find the correct ordering of the vertices of the convex hull. We can do this by merely examining the edges of any unbounded polygon Vor(v_i, T) until we find an edge that is shared by another unbounded polygon Vor(v_j, T). The vertices v_i and v_j are then consecutive vertices of CH(T). If we repeat the procedure, we can identify the order of the vertices of CH(T) in $O(|T|)$ steps.

Using CH(L) and CH(R) to Construct CH(L ∪ R) [(4.2)].   To construct CH(S) or CH($L \cup R$) from the pair of disjoint convex polygons CH(L) and CH(R), we need the following definition. A *supporting line* of a convex polygon Q is a straight line C that has at least one point in common with Q and is such that all the points of Q lie on one side of C. A pair of disjoint convex polygons have exactly two common supporting lines. They can be found in linear time. We leave the technique to the reader. The convex hull of $L \cup R$ can then be constructed using the common supporting lines of CH(L) and CH(R). Figure 2-6 gives an example. The line segments L_1 and L_2 in the figure are segments of the common supporting lines. They link pairs of vertices from the component convex hulls: L_1 links c to h and L_2 links e to j. The convex hull CH($L \cup R$) is then just CH(L) $\cup$ CH(R) $\cup L_1 \cup L_2$.

Using CH(L ∪ R) to Construct the Rays of P [(4.3)].   Once we have constructed CH($L \cup R$) or CH(S), the rays of P are merely the perpendicular bisectors of the supporting line segments L_1 and L_2.

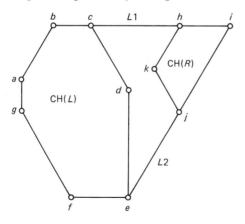

Figure 2-6.   CH($L \cup R$): Convex hull of a union of convex hulls.

Advancing P a Segment at a Time [(4.4)]. We find the remaining line segments of P one at a time, starting with the current segment of P. We begin with the ray of P that enters from the y direction minus infinity. Generally, we follow the current segment until we reach the next border of a polygon of Vor(L) or Vor(R). The next segment of P starts there and is perpendicular to the line segment between a pair of new points of S. We emphasize that the border just intersected is not itself the next P segment.

We now describe the procedure in more detail. Let D denote the selected starting ray. If D is orthogonal to the segment $L1$ running from v in L to w in R, then D is the border between the polygon Vor(v, S) and Vor(w, S). We proceed (from minus infinity) along D until we hit a border B of one of the next Voronoi polygons of Vor(L) or Vor(R) at a point r. B could be either a border between Vor(v, L) and a polygon Vor(v', L) for v' in L, or a border between Vor(w, R) and a polygon Vor(w', R) for w' in R. In the first case, the new P segment starts at r and follows the perpendicular bisector of v' and w. This segment is the border in Vor(S) between Vor(v', S) (a polygon to the left of P in Vor(S)) and Vor(w, S) (a polygon to the right of P in Vor(S)). We apply a similar procedure in the other case. We repeat the procedure until we reach the other ray of P.

Efficient Implementation of the Segment Advancing Technique. To find the segments of P for Vor(S) in $O(|S|)$ time, we first observe that

(P1) P is monotonically increasing in its second (y) coordinate, that is, P cuts any horizontal line only once.

(P2) A convex polygon is divided by its highest and lowest vertices into two sequences of line segments, each of which are separately monotonic in y, although perhaps not strictly so.

Let t denote the current segment of P. Let $t(y)$ denote the y ordinate where t started. Let Vor(v, L) and Vor(w, R) be the polygons P currently lies in. The next segment of P starts precisely when t meets the next border of either Vor(v, L) or Vor(w, R), whichever comes first.

We can find the border that triggers the next segment of P as follows. By (P2), the polygons Vor(v, L) and Vor(w, R) determine a total of four monotonic sequences M_i ($i = 1, \ldots, 4$) of line segments. Delete the lowest edge from each M_i until we come to an edge with the property that the y ordinate of its point of intersection with t is greater than $t(y)$. Whenever we examine an edge of a sequence M_i which t does not intersect above $t(y)$, we discard the edge since it cannot be the next segment of P. We then select from the four candidates (the first edges of each of the M_i) that edge A whose point of intersection with t has the smallest y ordinate. This is the first edge on any of the four edge sequences that t will intersect as it continues in its current direction.

After we identify A, we update the sequences M_i by identifying the next polygon of Vor(L) or Vor(R) that P will move through. This is determined by which edge of Vor(v, L) or Vor(w, R) t intersected. By P2, this polygon defines two new edge sequences that we use to replace the edge sequences of the polygon just exited.

The whole scanning process constructs P in linear time since none of the $O(N)$ edges of Vor(L) or Vor(R) is ever examined more than once.

Illustration of Construction of P. We will now illustrate the construction of P for an example (see Figures 2-7 and 2-8). Here, $S = \{v_1, v_2, v_3, v_4\}$, $L = \{v_1, v_2\}$, and $R = \{v_3, v_4\}$. Vor(L) consists of the single infinite ray B_{12} which perpendicularly bisects

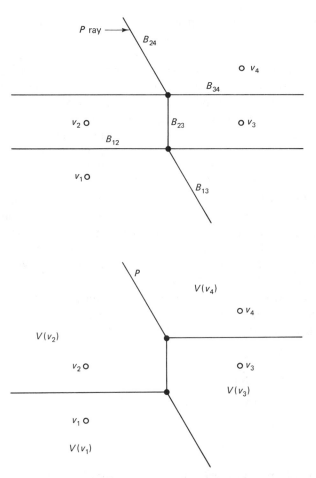
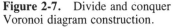

Figure 2-7. Divide and conquer Voronoi diagram construction.

Figure 2-8. Voronoi diagram on $\{v_1, v_2, v_3, v_4\}$.

the line segment (not shown) between v_1 and v_2. We will assume Vor(L) is available recursively. Similarly, for Vor(R), the ray B_{34}, the border with respect to Vor(v_3, R) and Vor(v_4, R), is also recursively available.

In order to find P from this information, we must first find the convex hull of $L \cup R$. In this example, the component convex hulls CH(L) and CH(R) are trivial. They are just the line segments v_1-v_2 and v_3-v_4. Their common support lines are the straight lines through v_1 and v_3 and through v_2 and v_4. Therefore, we can obtain the convex hull of their union by adding the edges (v_1, v_3) and (v_2, v_4), which we have not explicitly shown in Figure 2-7. The perpendicular bisectors of these two edges determine the two (entering and exiting) rays of the polygonal dividing boundary P which we seek. We denote the perpendicular bisector of (v_1, v_3) by B_{13}; while we denote the perpendicular bisector of (v_2, v_4) by B_{24}. B_{13} must be a border between Vor(v_1, S) and Vor(v_3, S) in the final Voronoi diagram Vor(S). Similarly, B_{24} must be a border between Vor(v_2, S) and Vor(v_4, S) in Vor(S).

To illustrate how to advance to the next segment of P, we begin with the ray B_{13}. It enters from infinity, along the common border of Vor(v_1, S) and Vor(v_3, S). As we move along this ray, we eventually intersect a border of either Vor(L) or Vor(R). Here, this occurs at B_{12}, a border of $V(L)$. Thereafter, at least until we intersect the next border of Vor(L) or Vor(R), P will be closer to v_2 than to v_1. Therefore, P now moves along the border B_{23} between Vor(v_2, S) and Vor(v_3, S), rather than the previous border between Vor(v_1, S) and Vor(v_3, S). Of course, since the points on this new border are, by the definition of P, equidistant from v_2 and v_3, this border must be just the perpendicular bisector of the line segment between v_2 and v_3, that is B_{23}. P continues to advance in this new manner until it intersects another edge of Vor(L) or Vor(R). In this example, this occurs when P reaches a point which is equidistant between v_3 and v_4, so that subsequently P will lie closer to v_4 than v_3. At this point P begins to proceed along the border between Vor(v_2, S) and Vor(v_4, S). Of course, this is again just the perpendicular bisector of the line segment B_{24}. In the example, this happens to be the exiting ray, and so this completes the construction of P.

Using P to Merge Vor(L) and Vor(R).   Once we have constructed P, the merging procedure is trivial. We merely superimpose the component Voronoi diagrams Vor(L) and Vor(R). We then eliminate the continuations of the edges of Vor(L) which lie to the right of P. Similarly, we eliminate the continuations of the edges of Vor(R) which lie to the left of P. Of course, both the bounded line segments and the rays of P are left as edges of Vor(S), the final Voronoi diagram.

Performance of Voronoi Algorithm.   The time $T(N)$ taken to construct the Voronoi diagram follows directly from its divide and conquer formulation and the linear time taken to merge the divided parts. $T(N)$ satisfies the recurrence relation

$$T(N) = 2T(N/2) + cN.$$

If we apply this recurrence repeatedly, we obtain

$$T(N) = 2^{\lg N} T(1) + c[N + 2(N/2) + \ldots + 2^{\lg N}(1)],$$

or, since the summation contains lg N terms equal to N,

$$T(N) = O(N \log N).$$

Nearest neighbor problem.   Given n points in the plane, how can one efficiently find the nearest neighbor of each? The obvious method is to compute all $O(n^2)$ intervertex distances and select the smallest at each vertex, which takes time $O(n^2)$. But, the Voronoi diagram can be used to solve the problem in $O(n \log n)$ time. We observe that the perpendicular bisector of any of the n points v_i and its nearest neighbor v_j is by definition just an edge of the Voronoi diagram. Therefore, to find the nearest neighbor of v_i, we merely examine each boundary edge of the Voronoi polygon containing v_i and select the one nearest v_i, which then determines v_j. Since each edge appears in exactly two Voronoi polygons, the algorithm examines each edge twice. Since there are $O(n)$ edges in the Voronoi diagram, the problem can thus be solved in linear additional time once the Voronoi diagram is available.

2-9 PROBLEM TRANSFORMATION

Perhaps the most classical problem solving technique is to reduce the solution of one problem to the solution of a previously solved problem or to a special case of the same problem. This approach is used extensively in the theory of computational complexity (see Chapter 10), where it is used to show the computational equivalence of certain problems. However, in those cases the efficiency of the reduction or problem transformation is secondary to the mere ability to implement the reduction in a polynomial amount of time. Transformation is a useful problem solving technique only if the transformation itself is efficient, and the transformed problem is easier to solve than the original problem.

It may happen that the simplification gained by a problem transformation is apparent only. For example, this is the case with the following specializing transformation of the minimum dominating set problem. Define a *dominating set* for a graph $G(V, E)$ as a set of vertices S in $V(G)$ such that every vertex in G either belongs to S or is adjacent to a member of S (see also Chapter 8). A related concept is a *bipartite dominating set* for a bigraph $G(V_1, V_2, E)$ which is defined as a set of vertices S in V_1 such that every vertex in V_2 is adjacent to at least one of the vertices in V_1. The following theorem shows that general minimum dominating sets are only as difficult to compute as bipartite minimum dominating sets.

Theorem (Domination to Bipartite Domination). Let $G(V, E)$ be a graph. Then there is a bipartite graph $B(V_1, V_2, E_b)$ of order $2|V(G)|$ associated with G, such that the size of the minimum dominating set of G equals the size of the minimum bipartite dominating set of B.

The proof is as follows. Denote the vertices of G by $v_1, \ldots, v_n$. Each vertex v_i has a neighborhood $N(v_i)$ consisting of v its neighbors. Denote the collection of such neighborhoods, some of which could coincide (!), over all vertices in G by N, and define a bipartite graph $B(V_1, V_2, E_b)$ where V_1 equals N, V_2 equals V, and there is an edge from $N(v_i)$ in V_1 to v_j in V_2 if and only if v_j lies in $N(v_i)$. There is a 1-1 correspondence between the dominating sets of G and the bipartite dominating sets of B. Thus, let $\{x_1, \ldots, x_k\}$ be a dominating set of G. Then, every vertex v in $V_2(B)$ is dominated by $N(x_i)$ where x_i dominates v in G. Conversely, suppose $\{N(y_1), \ldots, N(y_k)\}$ is a bipartite dominating set of B. Then, every vertex v in $V(G)$ is dominated by y_j where $N(y_j)$ dominates the copy of v in B. In particular, the cardinality of the smallest dominating set in G equals the cardinality of the smallest bipartite dominating set in B. This completes the proof.

The reduction in difficulty accomplished by the model of the theorem is only apparent, for the general problem and the bigraph version of the problem are both NP-Complete (see Chapter 10 for this terminology), so both are probably exponential. However, sometimes a transformation is truly useful, as the following example shows.

Maximum disjoint paths to 0-1 minimum perfect matching. Let $G(V, E)$ denote a graph and let X and Y be a pair of disjoint sets of vertices in $V(G)$. Define a pair of paths in G to be *pairwise vertex disjoint* if they have at most their endpoints in common. More generally, define a set of paths to be *vertex disjoint* if every pair of paths in the set is pairwise vertex disjoint. The maximum vertex disjoint paths problem is to find a set of vertex disjoint paths S such that each path in S has one endpoint in X and one in Y and S has maximum cardinality among all such sets. This problem can be solved using the flow techniques described in Chapter 6. We can also transform the problem into a minimum weight perfect matching problem on a transformed graph, which can be solved by an efficient parallel matching algorithm given in Chapter 9.

The procedure for constructing the transformed weighted graph $G'(V', E')$, with edge weights equal to 0 or 1 only, follows. For simplicity, we will assume that $|X| = |Y| = n$.

(1) Define $V'(G')$ as: for every vertex v in $V - X - Y$, create a pair of vertices $v(\text{in})$ and $v(\text{out})$. Denote the set of such vertices by W. Then, set V' to $W \cup X \cup Y$, where X and Y are copies of the sets in $V(G)$.

(2) Define $E(G')$ (or E') as: connect each $v(\text{in})$, $v(\text{out})$ pair by an edge $(v(\text{in}), v(\text{out}))$; also each edge (v, w) in $E(G)$ where v and w are in $V - X - Y$ gives rise to a pair of edges $(v(\text{in}), w(\text{out}))$ and $(w(\text{in}), v(\text{out}))$ in $E(G')$; each edge (x, v) with x in X and v in $V - X - Y$ in $E(G)$, gives rise to an edge $(x, v(\text{in}))$ in $E(G')$, while each edge (v, y) with y in Y and v in $V - X - Y$ gives rise to an edge $(v(\text{out}), y)$ in $E(G')$. For each edge (x, y) in $E(G)$, with x in X and y in Y, include the edge (x, y) in $E(G')$. All the edges defined thus far are assigned a weight of 0. Finally, we superimpose on G' all the edges induced by a complete bipartite graph with parts X and Y. These edges are assigned a weight of 1, and any parallel edges that result from their addition are retained! (Thus, G' may be a multigraph.)

We shall show there is a 1-1 correspondence between any set of k vertex disjoint paths from X to Y in G and a perfect matching of weight $n - k$ in G', and conversely. Therefore, we can find a maximum cardinality set of vertex disjoint paths from X to Y in G simply by finding a minimum weight perfect matching in G'. The minimum weight perfect matching has the effect of minimizing $n - k$, which is equivalent to maximizing k. There are efficient algorithms for finding minimum weight perfect matchings, so this problem transformation is useful. The theorem establishing the correspondence follows.

Theorem (Disjoint Paths to Perfect Matching Correspondence). Let $G(V, E)$ be a graph, and let X and Y be a pair of disjoint sets of vertices in G. Let $G'(V', E')$ be the 0-1 weighted graph obtained from G by the above procedure. Then, there is a 1-1 correspondence between the sets of k vertex disjoint paths from X to Y and the perfect matchings on G' of weight $n - k$.

The proof is as follows. First, let us show that for any set of k vertex disjoint paths in G there is a corresponding perfect matching of weight $n - k$ in G', which can be constructed as follows. Let $x_i, v_{i1}, \ldots, v_{im}, y_j$ be one of the disjoint X to Y paths. Let each of the edges $(x_i, v_{i1}(\text{in}))$, $(v_{im}(\text{out}), y_j)$, and $(v_{ir}(\text{out}), v_{ir+1}(\text{in}))$, for $r = 1, \ldots, m - 1$, in G' be matching edges. Any remaining unmatched vertices in W can be matched by $(v_i(\text{in}), v_i(\text{out}))$. Finally, match the remaining $n - k$ pairs of vertices of X and Y using the bipartite edges of weight one. The resulting matching is a perfect matching of weight $n - k$.

Conversely, any perfect matching M of weight $n - k$ in G' determines a set of k vertex disjoint paths from X to Y in G. Let M' be the set of edges in G' of the form $(v_i(\text{in}), v_i(\text{out}))$. Let H be the subgraph determined by the symmetric difference of M and M'. The vertices of X and Y in H all have degree one; while the vertices of W in H all have degree zero or two. Thus, H consists of only paths and cycles. The interior vertices of any path are alternately of the form $v_i(\text{in})$ and $v_j(\text{out})$; so the paths are from X to Y. There are k paths in H. Since the matching has weight $n - k$, k paths must have weight zero. These paths correspond directly to k vertex disjoint paths from X to Y in

G. This establishes the correspondence between the vertex disjoint paths of *G* and the perfect matchings of *G'*, and so completes the proof of the theorem.

Minimum weight maximum disjoint paths to minimum perfect matching. There is a generalization of the preceding problem to the case where the edges have weights, and the problem is to find a maximum set of vertex disjoint paths between *X* and *Y* of minimum total edge weight. We shall describe how this problem may be efficiently reduced to a minimum weight perfect matching problem, which can be efficiently solved by, for example, the parallel matching algorithm in Chapter 9.

We construct an associated graph $G''(k)$ as follows: the underlying construction is the same as that just given for *G'*. However, for *G''* any edges coming from *G* inherit the weights they had in *G*. Furthermore, instead of adding a complete bigraph between *X* and *Y*, we add two new sets of vertices X^* and Y^*, each of size equal to $n - k$, and then add complete bipartite graphs with edges of weight zero for *X* to X^*, as well as for *Y* to Y^*.

The following theorem establishes a correspondence between sets of *k* vertex disjoint paths of minimum weight on *G* and minimum weight perfect matchings on $G''(k)$.

Theorem (Min-weight Disjoint Paths to Minimum Weight Perfect Matching). Let $G(V, E)$ be a weighted graph, and let *X* and *Y* be a pair of disjoint sets of vertices in *G*. Let $G''(k)$ be the weighted graph associated with *G* by the above procedure. Then, there is a 1-1 correspondence between the sets of *k* vertex disjoint paths of minimum weight from *X* to *Y* in *G* and the minimum weight perfect matchings on $G''(k)$.

The proof is similar to the argument used in the preceding theorem. We merely prove there is a 1-1 correspondence between any set of *k* disjoint paths in *G* and a perfect matching on $G''(k)$, and that furthermore, the weight of the perfect matching equals the weight of the set of disjoint paths.

We can use this construction to find a maximum size minimum cost set of vertex disjoint paths in *G*. We merely apply the first construction for *G'* to find the size *k* of a maximum set of vertex disjoint paths between *X* and *Y* in *G*. Then, we apply the second construction for $G''(k)$ to find a minimum weight perfect matching on $G''(k)$, and hence, by the theorem, a minimum weight maximum set of vertex disjoint paths between *X* and *Y* on *G*.

2-10 INTEGER PROGRAMMING

Integer programming refers to the problem of optimizing, (maximizing or minimizing) a linear objective function, subject to a system of linear inequalities, which act as constraints on the integer variables over which the objective function is optimized. Integer programming is NP-Complete (refer to Chapter 10 for this terminology), and so probably has no polynomial algorithm. However, graph optimization problems can often be conveniently formulated as integer programming problems, which makes the various packages and techniques devised for integer programming available for the analysis of the problem. We will illustrate the modeling possibilities with a few examples.

Minimum vertex cover. A *minimum vertex cover* of a graph $G(V, E)$ is a set of vertices *S* in $V(G)$ such that every edge of *G* is incident with some vertex of *S*, and *S* has minimum possible cardinality among all such sets. (Refer to Chapter 8 for more on this topic.) The problem of finding a minimum vertex cover can be modeled as follows. Assume there is a 0-1 variable x_{ij} for every edge (i, j) in $E(G)$, and a 0-1 variable x_i for every vertex i in $V(G)$. Let $|V(G)|$ equal n. We will define a system of linear relations such that $(x_1, \ldots, x_n)$ is a solution of the system if and only if $S = \{i \mid x_i = 1\}$ defines a minimum cover of G, and that an edge variable x_{ij} equals 1 if and only if the edge (i, j) is incident with a vertex in S. The model is

$$0 \le x_{ij} \le 1, \quad \text{for every edge } (i, j), \tag{2-3}$$

$$0 \le x_i \le 1, \quad \text{for every vertex } i, \tag{2-4}$$

$$x_{ij} = x_{ji} \tag{2-5}$$

$$x_i \le x_{ij}, \qquad x_j \le x_{ij}, \tag{2-6}$$

$$x_{ij} \le x_i + x_j, \tag{2-7}$$

$$\sum_{i,j} x_{ij} = 2|E(G)|, \tag{2-8}$$

$$\min \sum_i x_i. \tag{2-9}$$

Inequalities (2-3) and (2-4) make the variables 0-1 variables, and the symmetry condition (2-5) merely ensures consistency. The inequalities (2-6) ensure edge (i, j) is covered if either endpoint i or j is covered; while conversely (2-7) ensures (i, j) is not covered if neither of its endpoints is covered. Inequality (2-8) requires every edge of G to be covered, while the objective function definition (2-9) ensures that the vertex cover found is of minimum cardinality.

Shortest path (Digraph). Given a digraph $G(V, E)$ with real weights assigned to its edges, we can model the shortest path problem between a pair of vertices s and t in G by a model similar to a linear programming model for the maximum flow problem (see Chapter 6). There are efficient algorithms for this problem (such as Ford's, see Chapter 3), so the integer programming model is primarily for the sake of illustration. Our objective is to constrain a set of edge variables x_{ij} so that the set of edges (i, j) for which x_{ij} is nonzero determines a shortest path between s and t. Denote the edge weights by w_{ij}. The integer programming model is

$$0 \le x_{ij} \le 1, \quad \text{for every edge } (i, j), \tag{2-10}$$

$$\sum_j x_{ij} \le 1, \ \sum_j x_{ji} \le 1, \quad \text{for every vertex } i, \tag{2-11}$$

$$\sum_j x_{ij} = \sum_j x_{ji}, \quad \text{for every vertex } i <> s, t \tag{2-12}$$

$$\sum_j x_{sj} = \sum_j x_{jt} = 1 \tag{2-13}$$

$$\sum_j x_{tj} = \sum_j x_{js} = 0 \tag{2-14}$$

$$\min \sum_i w_{ij} x_{ij}. \tag{2-15}$$

Inequality (2-10) ensures the 0-1 character of the edge variables. The graph is not directed, so there is no symmetry condition. The inequalities (2-11) ensure there is at most one nonzero edge into and out of any vertex; while (2-12) ensures there are an equal number of nonzero edges into and out of every vertex other than the initial vertex s and the terminal vertex t. The inequalities in (2-13) force there to be exactly one outgoing nonzero edge at s and exactly one incoming nonzero edge at t, while (2-14) forces there to be no nonzero incoming edge at s and none outgoing at t. The objective function (2-15) just sums the weights of the nonzero edges. The optimal solution defines a shortest s to t path, provided the graph satisfies the following condition: G contains no cycle whose total edge weight is negative. Otherwise, the optimal solution could correspond to a disjoint union of a path and a set of disjoint cycles, rather than to a shortest path as intended.

Shortest path (Undirected graph).   We now consider a model for an undirected version of the shortest path problem. The edge weights are allowed to be negative, but again there are no negative weight cycles. There is a polynomial time algorithm for this problem, but it is fairly complicated and is not given in Chapter 3. The integer programming model is as follows:

$$0 \le x_{ij} \le 1, \quad \text{for every edge } (i,j), \tag{2-16}$$

$$x_{ij} = x_{ji}, \quad \text{for every edge } (i,j), \tag{2-17}$$

$$\sum_j x_{ij} \le 2, \quad \text{for every vertex } i, \tag{2-18}$$

$$\sum_{j<>k} x_{ij} \ge x_{ik}, \quad \text{for each edge } (i,k), \ i <> s,t, \tag{2-19}$$

$$\sum_j x_{sj} = 1, \qquad \sum_j x_{jt} = 1, \tag{2-20}$$

$$\min \sum_i w_{ij} x_{ij}. \tag{2-21}$$

Inequalities (2-16), (2-17), and (2-21) have the usual interpretation. Inequalities (2-18) and (2-19) together imply any vertex other than s or t has either 0 or 2 nonzero incident edges. Inequality (2-20) makes s and t the endpoints of the path solution. The absence of cycles of negative weight, together with (2-18), (2-19), and (2-21), ensure the solution is a path of least weight.

2-11 PROBABILISTIC TECHNIQUES

Most graph-theoretic optimization problems are deterministic in nature, but despite this, random algorithms can sometimes provide elegant and efficient methods for their solution. This should not be surprising since random methods have long been used to evaluate deterministic processes. For example, Monte Carlo techniques are used to evaluate definite integrals. Several random algorithms are described in this book: the probabilistic algorithm for perfect matching in Chapter 1, and a randomized parallel algorithm for matching in Chapter 9; and a connectivity algorithm using a randomizing step is given in Chapter 6. Chapter 10 introduces a nomenclature for the complexity of different types of random algorithms, describes an efficient polynomial time algorithm for primality (a problem not known to be deterministically polynomial time), and gives a random distributed ringleader protocol algorithm.

We will now describe a fast random algorithm for the connected components of a graph due to Karp and Tarjan (1980). In contrast to the usual depth-first search algorithm for connected components, which has worst case performance $O(|V| + |E|)$, this algorithm has an *expected time* of only $O(|V|)$. The model of random graphs used assumes that every labelled graph $G(V, E)$ of fixed order $|V|$ and size $|E|$ is *equiprobable* (see Erdos and Renyi (1960)). The input to the algorithm is a randomly permuted list of the edges in $E(G)$, and an array of doubly linked adjacency lists whose sublists are also randomly permuted.

The idea of the method is to first find a large component of the input graph $G(V, E)$, that is, a component containing more than half the vertices in G. Erdos and Renyi (1960) showed there exists a *constant $c > 1/2$*, such that with probability p_n tending to 1 as n tends to infinity, a random graph of order and size n has such a "giant" component with at least cn vertices, so this step has a high probability of success. This is followed by a fix-up phase where we merge the remaining component fragments, either into the giant component, or recognize additional separate small components. The procedure has two steps.

(1) Find a "giant" component of $G(V, E)$.

> **Set** E' to empty
> **Set** n to |V(G)|
>
> **repeat**
>
>   **if** |E − E'| ≥ n
>
>   **then** Set A to a set of n distinct random edges from E − E'
>
>   **else** Set A to E − E'
>
>   **Set** E' to E' ∪ A
>
>   **until** E' = E **or** G(V,E') has a component of order ≥ cn

(2) Merge or identify the remaining components: Assuming (probabilistically) that a giant component of G is found, denote it by H, and label every vertex in H as "giant," and the remaining vertices of G as "unmarked." Then, repeat the following until every unmarked vertex in G has been reached. Perform a depth-first search starting at an unmarked vertex v. If a "giant" vertex is reached, terminate the search and label every vertex reached during the search as "giant"; otherwise, when the search is blocked before reaching the giant component, label every vertex reached as a member of a new component.

This procedure obviously eventually identifies the components of G; so it only remains to prove that each step can be implemented in $O(|V(G)|)$ expected time. The performance of step (1) is established by the following theorem.

Theorem (Linear Time Random Components). Let $G(V, E)$ be a graph. Then, the Tarjan-Karp random algorithm for finding the components of G takes $O(|V(G)|)$ expected time.

We only outline the proof which depends on the following lemmas.

Lemma 1 (Random Giant Component). Let $G(V, E)$ be a graph. Let p_n denote the probability that a random graph of order and size n has a component of order at least cn, where $c > 1/2$ is the Erdos-Renyi constant. Then, the expected time to find a (giant) component in G of order at least cn using step (1) of the procedure is

$$O\left(p_n \sum_{i \geq 1} ni(1 - p_n)^{i-1}\right),$$

which is $O(n)$.

The proof is simple. The algorithm succeeds in finding a giant component on the first try with probability p_n, after examining n edges. It succeeds on the second try with probability at least $p_n(1 - p_n)$, after examining $2n$ edges. In general, the algorithm succeeds on the kth try with probability at least $p_n(1 - p_n)^{i-1}$ after examining kn edges, whence the expression in the theorem. The $O(n)$ estimate follows by summing the expression and using the fact that p_n tends to 1 as n tends to infinity. This completes the proof of the lemma.

The proof of the performance of step (2) requires the following lemma, the proof of which is more difficult.

Lemma 2. Let H be the giant component in $G(V, E')$, and let v be a search vertex reached during step (2) not in H. If w in $ADJ(v)$ is selected in step (2), the probability w is in $V(H)$ given that (v, w) is not in E' is at least c.

We refer to Karp and Tarjan (1980) for the very technical proof of this lemma. To prove step (2) also has performance $O(n)$, we then argue as follows. When advancing the search tree from a vertex v, the depth first search procedure merely selects a previously unexplored vertex w from the randomized adjacency list $ADJ(v)$. By lemma 2, if (v, w) is not in E', the probability is at least c that w is in the giant component. Therefore, for any vertex v, the expected number of vertices in $ADJ(v)$ that must be examined before a vertex that terminates the step is found is at most $1/c$. Since the expected size of E' is $O(n)$, it follows that step (2) is also $O(n)$, as was to be shown. This completes the proof of the theorem.

REFERENCES AND FURTHER READING

Brassard and Bratley (1988) and Horowitz and Sahni (1978) present a thoroughly methodological approach to the design of algorithms. Brassard and Bratley includes an excellent introduction to probabilistic algorithms. Aho, Hopcroft, and Ullman (1974) also emphasizes design techniques, as well as algorithm analysis. Polya (1957) and Polya (1968) are classical studies of problem-solving techniques in mathematics, but the ideas are equally applicable to algorithm design. Hu (1982) is a delightful review of combinatorial algorithms, with an operations research flavor. Reingold, Nievergelt, and Deo (1977) reviews a variety of algorithmic techniques, and gives a particularly thorough discussion of backtracking. For a comprehensive study of the mathematical methods used in the analysis of algorithms, see also Purdom and Brown (1985). Of course, the works of Knuth, Volume 1 (1968, 1973), Volume 2 (1969, 1981), and Volume 3 (1973), represent a fundamental and unsurpassed treasure trove of results and problems on the design, analysis, and application of combinatorial algorithms and data structures. Sedgewick (1983) covers a fascinatingly broad array of algorithms, including basic numerical techniques, searching algorithms, string matching, and graph algorithms. See Hu (1970) for an exhaustive study of integer programming and networks flows.

Kernighan (1971) gives the dynamic programming algorithm for optimal program partitioning. Knuth (1971) and Melhorn (1975) consider optimal binary search trees. The pipe-organ arrangement for optimal record access is from Wong (1983). The discussion of backtracking is based on Hu (1982), and Reingold, et al. (1977), which is the source of the digraph isomorphism algorithm. Lin (1965) gives his approximate method for the traveling salesman problem. The matching approximation for the problem is due to Christofides (1976). Lawler, et al. (1984) reviews the problem. Johnson, et al. (1988) consider a new technique called "simulated annealing" for the solution of graph-theoretic problems, like the traveling salesman problem. Shamos (1975) and Shamos and Hoey (1975) introduce the Voronoi diagram for combinatorial purposes. The disjoint paths to matching transformation is given in Aggarwal and Anderson (1987). See Karp and Tarjan (1980) for a complete treatment of the linear expected time connectivity algorithm, and Erdos and Renyi (1960). The mixed backtracking-probabilistic algorithms in the exercises were suggested by Brassard and Bratley (1988).

EXERCISES

1. Suppose the vertex separator S for the partitioned shortest path algorithm has the property that the shortest paths between every pair of vertices in S lie only in S. Can this be used to improve the algorithm?

2. Can the following modified knapsack problem be solved by the kind of dynamic programming techniques given for the ordinary knapsack problem? The problem is to maximize

$$\sum c_1 x_1^2 \quad \text{subject to} \quad \sum w_i x_1^2 \leq b,$$

x_i integers.

3. Estimate the size of the backtrack tree for the n-queens problem using Monte Carlo estimation. Then, compare this estimate with the statistics for actual executions of the algorithm.

4. Examine the following randomized variation of the backtracking algorithm for the n-queens problem. Randomly place a queen in the next available column, subject only to the constraint that it does not attack a previously placed queen, repeating the process until a valid placement is obtained or the process is blocked. Estimate the probability of success, and the average number of columns examined in the case of failure. By iterating the algorithm until it succeeds, we obtain a so-called *Las Vegas algorithm* (see Chapter 10), which with probability 1 always finds a valid placement eventually. What are the average number of iterations required for success, and the average number of vertices of the backtrack tree examined? Compare these statistics with those for deterministic backtracking.

5. Combine the random and backtracking approaches to the n-queens problem by first using the method of the previous problem to determine a random initial configuration with a few queens, then extending this starting configuration using backtracking. Compare the performance of this method with direct backtracking.

6. Suppose n tasks are to be executed in parallel on k processors, where task x_i requires time t_i to execute. Assign the tasks to the processors so as to minimize the completion time of the overall system of tasks.

7. Let $G(V, E)$ be a graph with vertices numbered $1..|V|$; let G_i be the induced subgraph on vertices $1..i$; let S_i be the set of maximal complete subgraphs (or *cliques*) in G_i; and let x be in S_i. If $V(x)$ is in $\text{Adj}(i + 1)$, then $V(x) \cup \{i + 1\}$ induces a clique in S_{i+1}. Otherwise, x is a clique in S_{i+1}, and $(x \cap \text{Adj}(i + 1)) \cup \{i + 1\}$ induces a complete subgraph in G_i which may or may not be maximal. Prove every clique in S_{i+1} can be generated in this manner and design a corresponding clique generating algorithm.

8. Adapt the isomorphism algorithm to undirected graphs. Include heuristic tests for nonisomorphism, such as, nonidentity of degree sequence, before entering the backtrack phase. Contrast the performance of the resulting algorithm to the backtrack algorithm without the heuristic test. Is there an advantage in applying the heuristic test recursively? Explain and give an example.

9. Design a branch-and-bound backtracking algorithm to find a complete subgraph of maximum order in a graph. Modify the algorithm to find a complete subgraph of maximum weight in a weighted graph using branch-and-bound.

10. Verify the recursive relation for vertex connectivity

$$VC(G) = 1 + \min_{v \in V(G)} \{VC(G - v)\}.$$

for a few special graphs, and then give a proof of its correctness. Does a similar relation hold for edge connectivity? Give a proof or construct a counterexample.

11. Construct a sequence of examples where the greedy traveling salesman algorithm has increasingly poor behavior, unboundedly so in the sense of the relative error: the ratio of the absolute error to the optimal value.

12. The traveling salesman problem is computationally prohibitive to solve for large graphs. Given this, how can we properly test the performance of approximate algorithms for this problem, since the optimal solutions that presumably serve as reference points for comparison seem to be effectively undiscoverable?

13. Prove that if $\text{Chr}(G, x) = x(x - 1)^{n-1}$, where n equals the order of G, then G is a tree.

14. Experimentally examine Lin's method, both with respect to the time it takes and the relative errors of its answers.

15. Construct examples that have worst case error performance for both variations of the approximate traveling salesman algorithm; the eulerizing algorithm; and the perfect matching algorithm.

16. Implement Melhorn's greedy II algorithm. Compare the average cost of the trees it generates against the average cost of the trees generated by ordinary binary insertion.

17. Prove the Voronoi diagram on n points in the plane has $O(n)$ edges.

18. Implement Voronoi's algorithm. Then, use it to solve the nearest neighbor problem.

19. Determine the diameter of a set of n points in the plane in $O(n \log n)$ time using the Voronoi diagram Shamos (1975).

20. Implement the random connected components algorithm. This will require some knowledge of the method of depth first search from Chapter 5.

3

Shortest Paths

3-1 DIJKSTRA'S ALGORITHM: VERTEX TO VERTEX

A *weighted digraph* (or *network*) is a digraph $G(V, E)$ with real valued weights or lengths assigned to each edge. Equivalently, a weighted digraph is a triple (V, E, w) where V and E have the usual interpretation and w is a function that maps the elements of E into the reals. The *length* of a path in a weighted digraph is the sum of the lengths of the edges on the path. A *shortest path* between a pair of vertices x and y in a weighted digraph is a path from x to y of least length. The *distance* from x to y is defined as the length of a shortest path from x to y. The following model illustrates the application of shortest paths.

Model: visibility graphs. Consider the problem of finding shortest paths through a planar region strewn with obstacles. Specifically, let x and y be a pair of points in the plane, and suppose we want to find the shortest path between x and y that avoids the interiors of a set of polygonal regions R. We can model this as a shortest path problem on a weighted graph by introducing the *visibility graph* $G_R(V, E)$ associated with the configuration. Thus, let $Vert(R)$ denote the vertices of the polygonal obstacles of R. Then, set $V(G_R) = \text{Vert}(R) \cup \{x, y\}$. We will say a vertex u in $V(G_R)$ is *visible* from a vertex v in $V(G_R)$ if the line segment uv cuts the interiors of none of the regions of R. In the case that u is visible from v, we include the line segment (u, v) as an edge of $E(G_R)$, and let the euclidean length of the edge (u, v) be the weight of (u, v). Then, the shortest obstacle avoiding path from x to y is just the shortest path between x and y in G_R.

Dijkstra's method. Dijkstra's algorithm solves the following problem. Let $G(V, E)$ be a weighted digraph all of whose edge weights are positive, and let x and y be a pair of vertices in G. Find the shortest path from x to y in G and its length, or show there is none. The algorithm uses a search tree technique and is based on the observation that the kth nearest vertex to a given vertex x is the neighbor of one of the jth nearest vertices to x, for some $j < k$. It follows that if we can find the jth nearest vertex to x, for every $j < k$, then we can easily find the kth nearest vertex to x. For example, let $Near(j)$ denote the jth nearest vertex to x, let $Dist(u)$ denote the distance from x to any vertex u, and let $Length(u, w)$ denote the length of the edge from u to any neighbor w of u. Then, the kth nearest vertex to x is that vertex v that minimizes

$$Dist(Near(j)) + Length(Near(j), v),$$

where the minimum is taken over all $j < k$. Thus, to find the distance to y, we first inductively find the distances to all vertices closer to x than y.

The successively more distant vertices from x are found using a search procedure which explores the graph in a tree-like manner. This search induces a subdigraph of G called a *search tree*, which in turn contains a distinguished subtree called a *shortest path subtree* which has the property that the shortest distances to all its vertices are known. The search process is continued until the target vertex y is incorporated in the shortest path subtree.

At each phase of the search process, a new vertex v lying in the search tree but not in the shortest path subtree is incorporated in the shortest path subtree, and the search tree is then extended from v to its neighbors. For each such neighbor w of v, we make w point to v as its search tree predecessor if either w was not previously visited, or w was already visited, but the path through the tree from x to w is longer than the path through the tree from x to v plus the edge from v to w.

Initially, the search tree fans out from x to its immediate neighbors, defining a star at x. Observe that the second nearest vertex to x is necessarily that vertex in the star which lies at the end of a shortest edge incident with x. Trivial though this observation is, it already depends critically on the positivity of the edge weights. Furthermore, we can also discern at this point the distinction between the search tree, which includes all those vertices so far reached by the search procedure (which is currently just the star), and the shortest path subtree, consisting of x and its closest neighbor.

After k stages, the shortest path subtree of the search tree contains the k nearest vertices to x. The path through this tree from x to any of its vertices is a shortest path. On the other hand, the tree path from x to vertices in the search tree, which are not yet in the shortest path subtree, are estimated shortest paths only, and are subject to revision. We will show later that the $(k + 1)$-st nearest vertex to x is precisely that vertex v in the stage k search tree whose estimated distance from x is a minimum among all those vertices not yet in the shortest path tree.

An example search tree is shown in Figure 3-1. The edges labelled R (Reached) lead to vertices which are in the search tree, but not yet in the shortest path tree. The edges labelled S lead to vertices which are already in the shortest path subtree. Any vertices not shown correspond to vertices as yet unreached by the search. Any edges not shown are either edges as yet unexplored, or explored edges known to not lie on shortest paths.

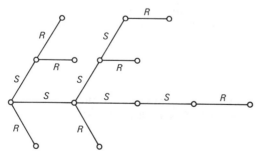

Figure 3-1. Search tree.

Dijkstra's algorithm.   In our statement of Dijkstra's algorithm, we will use the standard linear array representation for a graph (refer to Chapter 1). The definition of the type Vertex is slightly different.

```
type  Vertex = record
                  Dist: Real
                  Pred: 0..|V|
                  Positional Pointer: Edge pointer
                  Successor: Edge pointer
              end
```

The field Dist(v) will contain the current estimated distance to v from x calculated by the algorithm. Pred(v) gives the index of the search tree predecessor of v. The remaining fields in Vertex have the usual interpretation.

Although it is instructive to think of the algorithm in terms of a search tree, we do not need to explicitly maintain the tree as a separate data object. Instead, it is embedded in the predecessor pointers. Observe that although each search tree vertex has several search tree successor vertices, it has only a single search tree predecessor vertex; so a single predecessor pointer suffices at each vertex. After the algorithm finds the shortest path from a vertex x to a vertex y, we can read off the shortest path, in reverse order, by following the predecessor pointers from y back to x.

Although the algorithm does not explicitly maintain the search tree, we do explicitly maintain a set, Reached, consisting of vertices which already lie in the search tree, but are not yet in the shortest path subtree. We represent Reached in the algorithm as a set, but later on we will suggest a more efficient data structure for Reached.

Each edge in the graph has a defining record of the following form

```
type  Edge = record
                Length: Real
                Neighboring vertex: 1..|V|
                Successor: Edge pointer
            end
```

The field Length gives the length of the edge. Neighboring vertex identifies the other endpoint of the edge.

The function Dijkstra (G,x,y) either returns a shortest path from x to y, as embedded in the predecessor pointers starting at y, or fails because y is not reachable from x. The algorithm uses a function Getmin(v) which returns in v the vertex in Reached with the minimum value of Dist(v), and removes v from Reached, effectively placing v in the implicit shortest path subtree. Getmin fails if Reached is empty, and succeeds otherwise. The operator **and*** denotes conditional conjunction: the second operand is not evaluated if the first operand fails.

Function Dijkstra (G,x,y)

(* Returns the shortest distance from x to y in Dist(y), and
    the shortest path using the Pred fields starting at y, or
    fails. *)

```
var  G:  Graph
     v,w,x,y:  1..|V|
     M: Integer Constant
     Reached: Set of 1..|V|
     Dijkstra, Getmin: Boolean Function

Set Reached  to {x}
Set Pred(w)  to 0, for each vertex w in G
Set Dist(x)  to 0
Set Dist(w)  to M (Large), for each vertex w <> x in G

while  Getmin(v) and* v <> y  do

   for each neighbor w of v do

   if w  unreached

   then Add w to Reached
        Set Dist(w)  to Dist(v) + Length(v,w)
        Set Pred(w)  to v

   else if w in Reached  and  Dist(w) > Dist(v) + Length(v,w)

      then Set Dist(w)  to Dist(v) + Length(v,w)
           Set Pred(w)  to v

Set Dijkstra = ( v = y )

End_Function_Dijkstra
```

A weighted digraph G is shown in Figure 3-2. Figure 3-3 traces the development of the search tree for G rooted at v_1. The shortest path or distance to v_4 is sought. The shortest path tree edges are labelled S. The search tree edges are labelled S or R.

Theorem (Validity of Dijkstra Algorithm). If $G(V, E)$ is a network with non-negative edge weights, and x and y are in $V(G)$, and y is reachable from x, then Dijkstra's algorithm finds a shortest path from x to y.

We can prove the theorem by induction. Thus, assume that the first k iterations of the algorithm correctly identify the k nearest vertices to x. We will show that the $(k + 1)$st iteration correctly identifies the $(k + 1)$st nearest vertex to x.

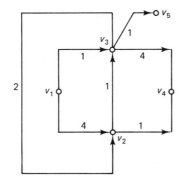

Figure 3-2. Weighted digraph G.

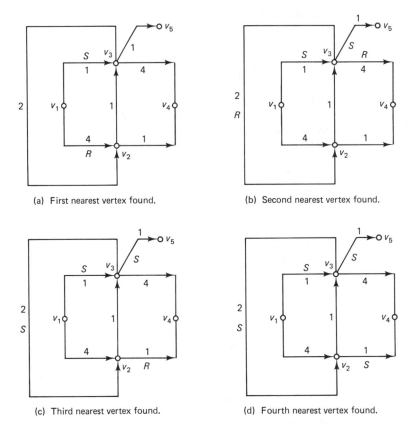

(a) First nearest vertex found.

(b) Second nearest vertex found.

(c) Third nearest vertex found.

(d) Fourth nearest vertex found.

Figure 3-3.   Trace of Dijkstra's algorithm for digraph of Figure 3-2.

Let y be the $(k + 1)$st nearest vertex to x. Let P be a shortest path from x to y. Let w be the predecessor of y on P. We assume that all edges have positive weight. The part of P from x to w is a shortest path; y lies one (positive) edge past the end of this path. Therefore, w must be closer to x than y, that is, w must be among the k nearest vertices to x. By induction, the algorithm correctly identifies a shortest path to w within the first k stages. Therefore, by the end of stage k, Dist(y) is set to Dist(w) + Length(w,y), which is the true shortest distance from x to y. By the beginning of stage $k + 1$ the algorithm recognizes y as the next nearest vertex to x and has already calculated its correct distance from x and a shortest path realizing that distance. This completes the proof by induction.

Analysis of Dijkstra's algorithm and enhancement.   The computationally critical steps in the algorithm are

(1) Finding the next nearest vertex v in Reached, and
(2) Updating the fields of the vertices adjacent to v.

Each step is iterated as many as $|V|$ times, since the target vertex may be as far as $|V|$ vertices away. The simplest way to implement step (1) is to linearly scan Reached to select the minimum. For step (2), as many as $O(|V|)$ neighbors of the selected vertex must be updated, taking $O(|V|)$ time. Accounting for iteration, the whole procedure then takes $O(|V|^2)$ steps.

We can try to refine this analysis by observing that there are only $\deg(v)$ neighbors at each selected minimum vertex v; so the total number of actions under (2) is at most $O(|E|)$. However, the total number of steps in (1) is still $O(|V|^2)$; so this yields no overall improvement in the performance estimate. Nonetheless, this analysis suggests (1) may be a bottleneck, so we may be able to improve performance by doing (1) more efficiently.

We can speed up (1) by representing Reached as a heap of vertices ordered on the value of their Dist field. (We only need to keep the index and value of Dist for each vertex in the heap.) Recall that a heap is a data structure that facilitates both sorting and minimum selection. We will assume familiarity with the definition of a heap. The basic *heap operations* are

(H1) Create (a heap),

(H2) Find_min (find the minimum element in the heap),

(H3) Delete (the least element in the heap and restore the heap),

(H4) Insert (a new element into a heap),

(H5) Member (test if an element is in a heap), and

(H6) Change (the value and if necessary the position of an existing heap element).

We will analyze the performance of heaps, particularly the Create and Delete operations, in detail in Section 4-3. Here, let us recall that we can delete the smallest element of a $|V|$ element heap in $O(\log|V|)$ time or add a new element to a heap in $O(\log |V|)$ time.

We can also Change the value of an existing heap element in $O(\log |V|)$ time, but this requires some care. To keep the Change operation efficient, we have to be able to directly access the heap element whose value is to be changed. We can do this by keeping an auxiliary array, Ptr, of pointers into the heap, where Ptr(i) gives the current position in the heap of the ith vertex of G. We can then implement Change much like an ordinary heap insertion. Thus, if the new changed value of an element is smaller than the value of its heap parent, the changed element moves up in the heap; otherwise, it is sifted down the heap in a standard manner (refer to Section 4-3).

If we represent Reached using a heap rather than a set, the operations (1), each of which entails a single heap delete operation, take $O(|V|\log |V|)$ time over all iterations. Each occurrence of step (2) requires $\deg(v)$ Insert or Change operations per selected vertex v. Each of these takes $O(\log |V|)$ steps, and so (2) requires $O(\deg(v)\log(|V|))$ steps or, summed over all iterations of (2), $O(|E| \log |V|)$ steps. Therefore, the total work is $O(|V| \log(|V|) + |E| \log |V|)$, which is just $O(|E| \log |V|)$. This is an improvement on the earlier estimate ($O(|V|^2)$) when $|E|$ is $O(|V|)$, but inferior if $|E|$ is $O(|V|^2)$.

Using the standard array representation for a heap, augmented to include information needed for our algorithm, we can represent a heap as follows:

```
type Vertex Heap = record
         H(|V|):   0..|V|
         Last:     0..|V|
         Ptr(|V|): 0..|V|
     end
```

The indices of the vertices in the heap are stored in H. We assume the heap is ordered on the Dist values of its vertices. In order to minimize redundancy, we keep these values only in the linear array representation of the graph G, where they are directly accessible from the vertex indices. Last gives the position in H of the last element of H. Ptr is the auxiliary array that provides direct access into H on the basis of the vertex index alone. It also lets us test for heap membership in $O(1)$ time by testing whether Ptr(w) is 0 or not.

Using this data type, we can restate Dijkstra's algorithm as follows. We will use the basic heap functions. Create (Reached) initializes the heap Reached. Insert (Reached, w) adds the vertex w to the heap Reached. Delete (Reached, v) deletes the smallest element of the heap, returning its index in v, and fails if the heap is empty. Member (Reached,w) succeeds if w is in Reached, and fails otherwise. Change (Reached, w) is used if Dist(w) is altered, and appropriately changes the position of w in the heap.

```
Function Dijkstra (G,x,y)

(* Returns the shortest distance from x to y in Dist(y), and
   the shortest path in Pred fields starting at y, or fails *)

var   G:  Graph
      v,w,x,y:  1..|V|
      M: Integer constant
      Reached:  Vertex Heap
      Delete, Member, Dijkstra: Boolean function

Create (Reached), Insert (Reached,x)
Set Pred(w)   to 0, for each vertex w in G
Set Dist(x)   to 0
Set Dist(w)   to M (large), for each vertex w <> x in G

while   Delete (Reached, v)  and*  v <> y  do

    for each neighbor w of v do

    if   w unreached

    then Set Dist(w)   to Dist(v) + Length(v,w)
         Set Pred(w)  to v
         Insert (Reached, w)

    else if Member(Reached,w)
               and*   Dist(w) > Dist(v) + Length(v,w)

         then Set Dist(w)   to Dist(v) + Length(v,w)
              Set Pred(w)  to v
              Change (Reached,w)

  Set Dijkstra = ( v = y )

  End_Function_Dijkstra
```

3-2 FLOYD'S ALGORITHM: VERTEX TO ALL VERTEX PAIRS

Dijkstra's algorithm requires the edge weights of a digraph $G(V, E)$ to be positive. Floyd's algorithm allows negative weights as well, but requires the digraph to be free of any cycle whose total edge weight is negative (a so-called *negative cycle*). The algorithm finds shortest paths between every pair of vertices in G and detects negative cycles if they occur. It also provides a compact matrix representation for the $|V|^2$ shortest paths found.

Floyd's method. The idea is to embed the original shortest paths problem in a parametrized sequence of subproblems, and then use induction (or dynamic programming) to solve the subproblems. The subproblems are defined as follows. Assume the vertices of $V(G)$ are indexed from $1..|V|$, and define an *internal vertex* of a path as any vertex on the path which is not an endpoint of the path. The kth subproblem (for $k = 0, \ldots, |V|$) is find the shortest paths between every pair of vertices in $G(V, E)$ where the internal vertices of the paths are chosen from vertices on $0..k$. For $k = 0$, there are no internal vertices on the paths, and so the shortest paths are just the given vertex to vertex edges of the input graph, where we follow the convention that a nonexistent edge is treated as if it were an edge of some large (infinite) weight. When $k = |V|$, the constraint is vacuous and the subproblem corresponds to the original problem.

At stage k, we have the shortest paths and distances between every pair of vertices, where the internal vertices have indices on $0..k$. We progress from the solutions at stage k to the solutions at stage $k + 1$ by allowing $k + 1$ as an intermediate vertex at stage $k + 1$ only if a stage $k + 1$ path through $k + 1$ improves the current shortest estimated distance. The decision is implemented by merely considering the alternative (use $k + 1$ or do not use $k + 1$) for every pair of vertices in G.

Let $P(i, j, k)$ denote a shortest path from i to j whose internal vertices have indices on $0..k$, and let $SD(i, j, k)$ denote the length of $P(i, j, k)$. Then

$$SD(i, j, k+1) = \min\{SD(i, j, k), SD(i, k+1, k) + SD(k+1, j, k)\}.$$

The validity of this procedure is less obvious than it seems, and we will establish it later. Refer to Figure 3-4 for an illustration. The rule merely says to always choose the better of the two possible ways of getting from i to j. The stage k path, $P(i, j, k)$, of length $SD(i, j, k)$ is the better path if $k + 1$ does not lie on a shortest stage $k + 1$ path

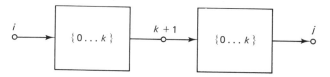

(a) Vertex $k + 1$ lies on optimal stage $k + 1$ path from i to j.

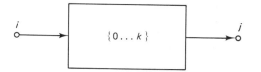

(b) Vertex $k + 1$ not on optimal stage $k + 1$ path from i to j.

Figure 3-4. Dynamic programming alternatives.

from i to j; while the path through $k + 1$ and of length $SD(i, k+1, k) + SD(k+1, j, k)$ is the better path if $k + 1$ does lie on a stage $k + 1$ shortest path from i to j.

Floyd's algorithm. The statement of the algorithm is simple. We represent the graph by its distance matrix *Dist*, where $Dist(i, j)$ gives the length of the (i, j) edge, where diagonal entries are set to zero, and if there is no edge between i and j (for $i <> j$), $Dist(i, j)$ is set to some large positive integer M. The type graph G is

```
type   Graph = record
               |V|: Integer Constant
               Dist(|V|,|V|)
          end
```

We store the stage k shortest distances in a $|\mathbf{V}| \times |\mathbf{V}| \times (|\mathbf{V}| + 1)$ array $SD(i, j, k)$. Later, we will observe that the algorithm can be done in place. The ordering of the loop indices in the algorithm is important. The outermost of the nested for loops must be indexed by the stage k. This ensures all the stage $k - 1$ values are available before the stage k computation needs them.

```
Procedure Floyd (G,SD)

(* Returns shortest distance between i and j in SD(i,j,|V|) *)

var   G: Graph
      i,j,k: 1..|V|
      SD(1..|V|, 1..|V|, 0..|V|): Real

for i, j = 1..|V|  do   Set SD(i,j,0) to Dist(i,j)

for k = 1 .. |V| do
 for i = 1 .. |V|  do
   for j = 1 .. |V|  do
     SD(i,j,k) = min {SD(i,j,k − 1),SD(i,k,k − 1) + SD(k,j,k − 1)}

End_Procedure_Floyd
```

The algorithm as given does not calculate the shortest paths, but a simple refinement lets us both calculate them and store them very compactly. Define $SP(i, j, k)$ as the successor of vertex i on the stage k shortest path from vertex i to vertex j. See Figure 3-5. For example, in Figure 3-6, $SP(1, 2, 4)$ is 3, because the stage 4 shortest path from v_1 to

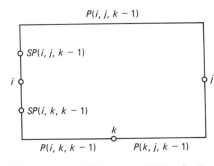

Figure 3-5. $SP(i, j, k) = SP(i, j, k-1)$ or $SP(i, k, k-1)$.

Figure 3-6. $SP(1, 2, 4) = 3$.

Shortest Paths   Chap. 3

v_2 is v_1-v_3-v_4-v_2. $SP(*, *, |V|)$ stores in implicit form the $|V|^2$ shortest paths between the $|V|^2$ pairs of vertices of G. This economical representation of $|V|^2$ paths is interesting in itself. Since there are $|V|^2$ paths, each containing as many as $O(|V|)$ vertices, it seems that we need $O(|V|^3)$ storage to represent the paths. SP represents the paths in $O(|V|^2)$ storage, though at the expense of $O(|V|)$ processing time to extract the vertices of a particular path.

We can calculate SP by a simple modification of Floyd's algorithm. If $P(i,j,k)$ contains k, the part of $P(i,j,k)$ from i to k is just $P(i,k,k-1)$. In that case, the successor of i on $P(i,j,k)$ (namely $SP(i,j,k)$) is the same as the successor of i on $P(i,k,k-1)$ (namely $SP(i,k,k-1)$). If $P(i,j,k)$ does not contain k, then $S(i,j,k-1)$ does not change. The modified procedure follows.

Procedure Floyd_Paths (G,SD,SP)

```
(* Returns the shortest distance i to j distance in SD(i,j,|V|)
   and the shortest paths between all pairs of vertices in G in
   SP(*,*,|V|) *)

var   G: Graph
      i,j,k: 1..|V|
      SD(1..|V|, 1..|V|, 0..|V|): Real
      SP(1..|V|, 1..|V|, 0..|V|): 1..|V|

for i, j = 1..|V|  do:   Set SD(i,j,0) to Dist(i,j)
                         Set SP(i,j,0) to j

for k = 1 .. |V| do
for i = 1 .. |V|  do
for j = 1 .. |V| do

    if      SD(i,j,k − 1) < SD(i,k,k − 1) + SD(k,j,k − 1)

    then    Set SD(i,j,k) to SD(i,j,k − 1)
            Set SP(i,j,k) to SP(i,j,k − 1)

    else    Set SD(i,j,k) to SD(i,k,k − 1) + SD(k,j,k − 1)
            Set SP(i,j,k) to SP(i,k,k − 1)
```

End_Procedure_ Floyd_Paths

We can use the procedure Extract_Shortest_Path to extract the shortest paths from SP. The procedure uses an auto-incrementing integer function Nextcount which is initially 1. P is initially identically 0.

Procedure Extract_Shortest_Path (SP, |V|, i, j, P)

```
(* Returns the shortest path from i to j in P *)

var   |V|: Integer Constant
      SP(1..|V|, 1..|V|,0..|V|): 1..|V|
      P(|V|): 0..|V|
      i, j, Next: 1..|V|
      Nextcount: Integer function
```

Set P(Nextcount) to i
Set k to i

while k <> j **do**   **Set** k to SP(k, j, |V|)
                        Set P(Nextcount) to k

End_Procedure_Extract_Shortest_Path

Performance of Floyd's algorithm.   The time performance of the algorithm is $O(|V|^3)$. For clarity, we have separated the data from different stages so that the space requirements are apparently $O(|V|^3)$. However, the values at stage k obviously depend only on the values from stage $k - 1$; so we actually require only $O(|V|^2)$ memory (such as two $|V| \times |V|$ arrays whose roles alternate.) Indeed, the computations can even be done in place. Observe that SD(i,j,k) depends on SD(i,j,k−1), SD(i,k,k−1), and SD(k,j,k−1). The last two terms remain constant at stage k, because vertex k is not an internal vertex on paths from i to k or k to j. Furthermore, no other stage k term except SD(i,j,k) depends on SD(i,j,k−1). Consequently, the storage for SD(i,j,k) and SD(i,j,k−1) can be overlaid. Thus, the computations can be done in place, and the algorithm needs to make no lexical distinction between the terms from the various stages. We can *rewrite floyd's algorithm in an in-place form* by merely replacing every occurrence of SD(i,j,k) and SP(i,j,k) in Floyd or Floyd_paths by an occurrence of SD(i,j) and SP(i,j).

An example of the algorithm is shown in Figures 3-7 and 3-8. Recall that neither row nor column k changes at stage k, and the diagonal entries never change unless the digraph has a negative cycle. When the graph contains negative cycles, an entry $SD(i, i)$ will eventually become negative if i lies on such a cycle. If this occurs it indicates the graph fails to satisfy the conditions of applicability of Floyd's algorithm. The graph in Figure 3-7 has no negative cycles. The initial **SD** and **SP** matrices are shown in Figure 3-8a. The first iteration allows v_1 as an intermediate vertex, with the results shown in Figure 3-8b. We have starred SD values that change at a given stage.

Validity of Floyd's algorithm.   The correctness of the algorithm depends on the following structural characteristics of shortest paths in digraphs with no negative cycles.

(1) *Path Optimality:* The shortest walk between a pair of vertices is always reducible to an equal valued shortest path.

(2) *Subpath Optimality:* The subpaths of a shortest path are also shortest paths.

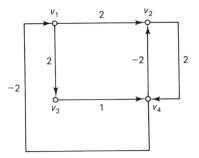

Figure 3-7. Weighted graph G.

$$
\begin{array}{cccc}
0 & 2 & 2 & M \\
M & 0 & M & 2 \\
M & M & 0 & 1 \\
-2 & -2 & M & 0
\end{array}
\qquad
\begin{array}{cccc}
1 & 2 & 3 & 4 \\
1 & 2 & 3 & 4 \\
1 & 2 & 3 & 4 \\
1 & 2 & 3 & 4
\end{array}
$$

(a) Initial **SD** Matrix.   (a') Initial **SP** Matrix

$$
\begin{array}{cccc}
0 & 2 & 2 & M \\
M & 0 & M & 2 \\
M & M & 0 & 1 \\
-2 & -2 & 0^* & 0
\end{array}
\qquad
\begin{array}{cccc}
1 & 2 & 3 & 4 \\
1 & 2 & 3 & 4 \\
1 & 2 & 3 & 4 \\
1 & 2 & 1^* & 4
\end{array}
$$

(b) **SD** after stage 1.   (b') **SP** after stage 1.

$$
\begin{array}{cccc}
0 & 2 & 2 & 4^* \\
M & 0 & M & 2 \\
M & M & 0 & 1 \\
-2 & -2 & 0 & 0
\end{array}
\qquad
\begin{array}{cccc}
1 & 2 & 3 & 2^* \\
1 & 2 & 3 & 4 \\
1 & 2 & 3 & 4 \\
1 & 2 & 1 & 4
\end{array}
$$

(c) **SD** after stage 2.   (c') **SP** after stage 2.

$$
\begin{array}{cccc}
0 & 2 & 2 & 3^* \\
M & 0 & M & 2 \\
M & M & 0 & 1 \\
-2 & -2 & 0 & 0
\end{array}
\qquad
\begin{array}{cccc}
1 & 2 & 3 & 3^* \\
1 & 2 & 3 & 4 \\
1 & 2 & 3 & 4 \\
1 & 2 & 1 & 4
\end{array}
$$

(d) **SD** after stage 3.   (d') **SP** after stage 3.

$$
\begin{array}{cccc}
0 & 1^* & 2 & 3 \\
0^* & 0 & 2^* & 2 \\
-1^* & -1^* & 0 & 1 \\
-2 & -2 & 0 & 0
\end{array}
\qquad
\begin{array}{cccc}
1 & 3^* & 3 & 3 \\
4^* & 2 & 4^* & 4 \\
4^* & 4^* & 3 & 4 \\
1 & 2 & 1 & 4
\end{array}
$$

Figure 3-8. Trace of **SD** and **SP** for G.

(e) **SD** after stage 4.   (e') **SP** after stage 4.

Path Optimality can fail if the graph has negative cycles. Figure 3-9 shows a graph containing a negative cycle where the shortest walk v_1-v_3-v_2-v_3-v_4 has length 1, while the two shortest paths v_1-v_3-v_4 and v_1-v_2-v_3-v_4 have length 2. For this graph, the shortest walk is shorter than the corresponding shortest path. Figure 3-10 shows a graph with a shortest walk v_1-v_3-v_2-v_3-v_4 which has the same length as the shortest path v_1-v_3-v_4 obtained by eliminating the zero weight cycle v_3-v_2-v_3 from the walk.

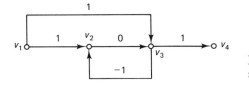

Figure 3-9. Shortest v_1-v_4 walk $<$ shortest v_1-v_4 path.

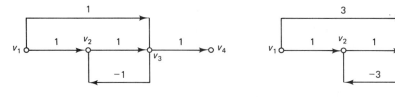

Figure 3-10. Shortest v_1-v_4 walk = shortest v_1-v_4 path.

Figure 3-11. Subpath property fails.

Subpath Optimality can also fail if the graph has a negative cycle. Figure 3-11 shows such a graph; v_1-v_2-v_3 is a shortest path (walk) from v_1 to v_3, but v_1-v_2 is not a shortest path from v_1 to v_2 (v_1-v_3-v_2 is). However, if the graph contains no negative cycles, we can prove the following theorem.

Theorem (Path and Subpath Optimality). Let $G(V, E)$ be a weighted digraph containing no negative cycle. Then, G satisfies both path and subpath optimality.

To prove path optimality holds, we argue as follows. Observe that a walk can always be reduced to a path by iteratively removing cycles from the walk. The process is concretely illustrated in Figure 3-12. Figures 3-12a and 12b show a walk from v_1 to v_5 with the multiply traversed edge v_2-v_3. The walk is v_1-v_2-v_3-v_4-v_2-v_3-v_5. Figure 3-12c shows the walk with the cycle v_2-v_3-v_4-v_2 eliminated. Each multiple occurrence of an edge on the cycle is removed from the walk. The resulting reduced walk is a path. The same process works in general. Consequently, if every cycle has nonnegative weight, every walk can be reduced to a path of at most equal weight. Therefore, a shortest walk can be reduced to an equivalent shortest path.

To prove subpath optimality we argue as follows. Refer to Figure 3-13 for an illustration of the following notation. Let $P(i, j)$ be a shortest path from i to j. Suppose k

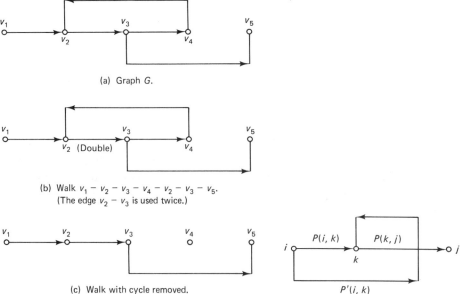

(a) Graph G.

(b) Walk $v_1 - v_2 - v_3 - v_4 - v_2 - v_3 - v_5$.
(The edge $v_2 - v_3$ is used twice.)

(c) Walk with cycle removed.

Figure 3-12. Reducing a walk to a path.

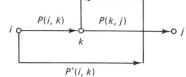

$P'(i, k)$

Figure 3-13. Subpath optimality.

lies on $P(i,j)$. Then $P(i,k)$ must be a shortest path from i to k. Otherwise, there exists a still shorter path $P'(i,k)$ from i to k. The union $P'(i,k) \cup P(k,j)$ is a walk from i to j. (It might not be a path, as illustrated in Figure 3-13.) By path optimality such a walk can be reduced to an equal valued path from i to j. But, since $P'(i,k)$ is shorter than $P(i,k)$, this path would be shorter than the presumed shortest path $P(i,j)$ from i to j. It follows by contradiction that $P(i,k)$ is a shortest path from i to k. This completes the proof of the theorem.

Theorem (Correctness of Floyd's Algorithm). Let $G(V,E)$ be a weighted digraph containing no negative cycle. Then, Floyd's algorithm correctly finds all inter-vertex distances.

The proof follows easily from path and subpath optimality. Thus, the relation

$$\mathbf{SD}(i,j,k+1) = \min \{\mathbf{SD}(i,j,k), \mathbf{SD}(i,k+1,k) + \mathbf{SD}(k+1,j,k)\}$$

merely asserts that a stage $k + 1$ shortest path either contains $k + 1$ or does not. If it does not, the optimal stage $k + 1$ solution is the same as the optimal stage k solution. If it does, then by subpath optimality the two subpaths used at stage k are shortest paths whose values were correctly calculated by stage k. The optimal stage $k + 1$ path can be a combination of two such paths or a stage k path. By path optimality, we know the combination of the two subpaths is a path. The algorithm determines the correct decision by selecting the best of the two alternatives.

3-3 FORD'S ALGORITHM: VERTEX TO ALL VERTICES

Ford's algorithm finds the shortest paths from a given vertex v to every vertex reachable from v in a weighted digraph $G(V,E)$ with real edge weights and containing no negative cycles. When applied to a digraph containing negative cycles, the algorithm will fail to terminate within $|V| - 1$ iterations, confirming the presence of a negative cycle.

Ford's algorithm solves the original problem by effectively embedding it in a sequence of subproblems, like Floyd's algorithm, although in this case the subproblems are only implicit. The embedding parameter is the number of edges on the path, and the algorithm finds successively longer shortest paths, that is, shortest paths emanating from v with increasing numbers of edges. By the end of its kth iteration, the algorithm finds all the shortest paths emanating from v that have at most k edges.

We can also visualize Ford's algorithm as constructing a search tree, like Dijkstra's algorithm. However, the search tree is more volatile than the search tree in Dijkstra's algorithm. As usual, we maintain a predecessor pointer, for each vertex u, that points to the predecessor of u on the current best shortest path $P(u)$ from v to u. The algorithm scans the edge list repeatedly, and, whenever adding an edge (w,u) to an estimated shortest path $P(w)$ leads to a shorter path to u, it makes w the predecessor of u. The algorithm is stopped at any iteration that yields no improvement in any of the estimated distances, or after $|V| - 1$ iterations, whichever comes first.

Ford's algorithm. We use an edge list representation for $G(V, E)$.

```
type  Graph = record
                |V|: Integer Constant
                |E|: Integer Constant
                Edges(|E|,2): 1..|V|
                Length(|E|): Real
                Dist(|V|): Real
                Pred(|V|): 0..|V|
      end
```

Edges stores the edge list for G, the pairs ($Edges(i, 1)$, $Edges(i, 2)$) being the ith edge of G, $i = 1 .. |E|$. $Length(i)$ gives the length of the ith edge, $i = 1 .. |E|$. During execution, $Dist(i)$ equals the length of the estimated shortest path to vertex i, and $Pred(i)$ gives the predecessor of i on the path.

We define the procedure Ford(G,v) to return in Dist(u) the shortest distance from v to u. The shortest paths can be extracted by following the predecessor pointers from a given vertex back to v. If a vertex is unreachable from v, Dist is left equal to the large default value M. The procedure fails if changes occur in the distance estimates during the $|V|^{\text{th}}$ pass, which can happen only if G has a negative cycle.

```
Function Ford (G,v)

(* Finds the shortest paths from v to every vertex reachable from
   v, and fails if it reaches a negative cycle *)

var  G: Graph
     Pass: 0..|V| − 1
     u,v,w: 1..|V|
     M: Integer Constant
     Ford: Boolean function

Set Pass   to 0
Set Dist(v) to 0
Set Dist(u) to M,   for u <> v, for M large

repeat

   Set Ford to True
   Set Pass to Pass + 1

   for every edge (u,w) in G do

      if   Dist(u) + Length(u,w) < Dist(w)

      then Set Dist(w)  to Dist(u) + Length(u,w)
           Set Pred(w) to u
           Set Ford    to False

   until  Ford  or*  Pass ≥ |V|

End_Function_Ford
```

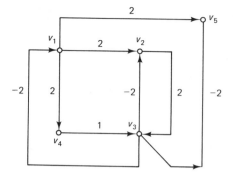

Figure 3-14. Weighted graph G.

An example is shown in Figure 3-14 and worked out below.. If an edge (u, w) improves the estimated distance to w, we star the revised entries.

	Edge(u, w)	Pass 1		Pass 2		Pass 3	
		Dist(w)	Pred(w)	Dist(w)	Pred(w)	Dist(w)	Pred(w)
(1)	(1, 2)	2*	1*	2	1	1	3
(2)	(1, 4)	2*	1*	2	1	2	1
(3)	(1, 5)	2*	1*	2	1	1	3
(4)	(2, 3)	4*	2*	3	2	3	4
(5)	(3, 1)	0	0	0	0	0	0
(6)	(3, 2)	2	1	1*	3*	1	3
(7)	(3, 5)	2	1	1*	3*	1	3
(8)	(4, 3)	3*	4*	3	4	3	4

Theorem (Correctness of Ford's Algorithm). Let $G(V, E)$ be a weighted digraph with real edge weights and containing no cycles of negative weight. Let v be in $V(G)$. Then, Ford's algorithm finds the shortest distances from v to every other vertex in G reachable from v.

The proof is by an induction on the number of edges in a shortest path. The proof has elements in common with the proofs of both Dijkstra's and Floyd's algorithm. We will rely on properties of shortest paths in digraph's containing no negative cycles described in the proof for Floyd's algorithm. They were path optimality, the shortest walk from v to any other vertex u is a path; and subpath optimality, the leading subpaths of a shortest path are shortest paths. Subpath optimality implies that if P is a shortest path from v to u and w is a predecessor of u on that shortest path, the subpath of P from v to w is a shortest path from v to w. Path optimality implies that the shortest walk from, say, v to u can always be chosen to be a shortest path from v to u.

At pass 1, the algorithm correctly identifies shortest paths emanating from v containing one edge. Assume by induction that by pass $k - 1$ the algorithm has determined the shortest paths from v containing $k - 1$ edges. (Observe that even though the correct $k - 1$ edge shortest paths have been determined, we do not actually recognize them until the algorithm terminates, which may be many passes later). Let u be a vertex such that the shortest path from v to u, denoted by $P(u)$, contains k edges. Let w be the predecessor of u on $P(u)$. By subpath optimality, the part of $P(u)$ from v to w is a shortest path from v to w, containing $k - 1$ edges. By induction, this subpath is correctly found before the kth iteration. Therefore, by the end of the kth iteration, $Dist(u)$ has received its correct value, when the edge (w, u) is examined. Since this distance estimate cannot be improved upon, it should not be changed in subsequent passes of the algorithm. Since the only phenomenon that could cause the algorithm to blindly make such a change (and so violate the shortest path interpretation of the estimate) would be due to the effect of a negative cycle, and since these are prohibited, the estimate remains at this optimal value. Consequently, the algorithm correctly calculates the distance to every vertex reachable from v. This completes the proof of the theorem.

Theorem (Performance of Ford's Algorithm). Let $G(V, E)$ be a weighted digraph containing no cycles of negative weight. Let v be in $V(G)$. Then, Ford's algorithm finds the shortest path from v to every vertex reachable from v in time $O(|V||E|)$.

The proof is trivial. The time performance of the algorithm is $O(|V||E|)$ because, at most, $|V|$ passes are required and each pass examines at most $|E|$ edges.

Ford's algorithm differs from both Dijkstra's and Floyd's algorithms in one interesting respect. If Dijkstra's algorithm is terminated at some stage k before the target vertex is reached, the k nearest vertices found by the algorithm to that point are identifiable from the partial results provided by the algorithm. Similarly, if Floyd's algorithm is terminated at the kth stage, then shortest paths using only the first k vertices internally are identifiable. In contrast, if Ford's algorithm is terminated at the kth stage, although one may know that all cardinality k shortest paths have been found by that point, one can nonetheless not tell which of the distance estimates are final and which are only tentative.

Effect of negative cycles. The algorithm may or may not operate correctly when the digraph contains negative cycles. It depends on the characteristics of the cycles.

Figure 3-15 shows a digraph containing a negative cycle which is unreachable from v_1. If we apply Ford's algorithm to this graph, with v_1 as the starting vertex, it will correctly calculate all shortest paths starting at v_1. Since none of the vertices on the negative cycle are reachable from v_1, none of the vertices on the negative cycle have their *Dist* values changed from their initial states. Thus, the algorithm terminates correctly.

Figure 3-16, on the other hand, shows the kind of negative cycle that makes the algorithm fail. In this example, and regardless of the order in which the edges are scanned, the distances to v_2 and v_3 are correctly estimated after two iterations. However, the algorithm assigns increasingly lower estimated distances to v_4, v_5, and v_6 as the iterations proceed. These estimates do not correspond to upper bounds on the lengths of shortest paths, as is the case when there are no negative cycles. But, they do have a graphical interpretation. They correspond to increasingly shorter walks from v_1 to vertices on the negative cycle. Not only are the estimated distances for the cycle ver-

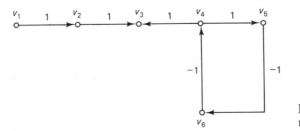

Figure 3-15. Negative cycle unreachable from v_1.

Shortest Paths   Chap. 3

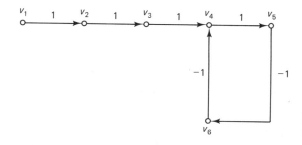

Figure 3-16. Negative cycle reachable from v_1.

tices affected, the distance estimates for any vertices reachable from such a negative cycle are also contaminated. Nonetheless, even under these circumstances the algorithm correctly calculates the shortest distances to vertices reachable from v_1 which are not reachable from any of the negative cycles.

3-4 EUCLIDEAN SHORTEST PATHS: SEDGEWICK-VITTER HEURISTIC

A *euclidean graph* $G(V, E)$ is a weighted graph embedded in euclidean n-space whose vertices correspond to points in n-space and in which the weight of an edge (u, v) is equal to the euclidean distance between the points u and v. Observe that in the case that G is drawn in the plane, G need not be planar. We will describe a heuristic modification of Dijkstra's algorithm for euclidean graphs that finds shortest paths in better expected (average) time than Dijkstra's algorithm. The worst case performance remains the same.

Let us recall the structure of Dijkstra's algorithm for finding the shortest distance through a directed graph $G(V, E)$ from a given vertex x to a target vertex y. The method extends a search tree T containing a shortest path subtree S until S reaches y. During the search process, the vertices are subdivided into three classes depending on whether they are unreached (not yet in T), reached (in T, but not yet in S), or lie in the shortest path tree S.

The Sedgewick-Vitter algorithm adaption of Dijkstra's algorithm is as follows. We maintain, for each vertex w in Reached, a lower bound, denoted by Heur_Dist(w), on the distance through G from x to y on a shortest path through w. Heur_Dist(w) is defined as the length of the path from x to w through the search tree plus the euclidean distance from w to y. While Dijkstra's algorithm uses the distance through the search tree Dist(u) to determine which vertex u to select as the next vertex to enter S; the Sedgewick-Vitter algorithm selects as the next vertex to enter S that vertex u that minimizes Heur_Dist(u). When the search reaches y, the algorithm terminates with the shortest path to y.

Sedgewick and Vitter have analyzed the algorithm for various classes of random euclidean graphs and shown its average performance is $O(|V|)$, provided the set of reached vertices is implemented as a so-called Fibonacci heap (see Fredman and Tarjan [1987]). In contrast, using a similar implementation, Dijkstra's algorithm attains a worst case performance of $O(|E| + |V| \log |V|)$. We refer the reader to Section 3-5 for a detailed treatment of Fibonacci heaps.

There are two reasons for the superior performance of the Sedgewick-Vitter algorithm. First, the algorithm terminates as soon as y is reached. Second, the heuristic distance estimates used to select the next neighbor to enter S are designed to tend to make

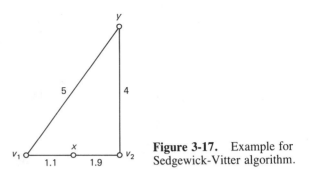

Figure 3-17. Example for Sedgewick-Vitter algorithm.

the search tree grow in the direction of y. This is only a tendency and not a strictly guaranteed effect.

Refer to Figure 3-17 for an example. When the search tree Reached is first extended from x, Heur_Dist(v_1) is 6.1, while Heur_Dist(v_2) is only 5.9. Thus, the first vertex added to S after x is v_2, even though v_1 is closer to x. Consequently, the shortest path to y will be identified on the next iteration of the algorithm, in contrast to Dijkstra's algorithm where all the vertices closer to x than y are finalized before the shortest distance to y is determined.

The algorithm follows. The utilities and data structures are similar to those for Dijkstra's algorithm. An additional field Heur_Dist is included for each vertex. Reached is implemented as a heap which is ordered with respect to Heur_Dist.

Procedure Sedgewick_Vitter (G,x,y)

(* Find the shortest distance from x to y in G *)

var G: Graph
v,w,x,y: 1..|V|
M: Integer constant
Reached: Vertex Heap
Delete, Member, Dijkstra: Boolean function

Set Pred(w) to 0, for each vertex w in V(G)
Set Dist(x) to 0
Set Dist(w) to M (large), for each vertex w <> x in G
Set Heur_Dist(x) to Euclid_Dist(x,y)
Create (Reached), Insert (Reached,x)

while Delete (Reached, v) **do**

  for each neighbor w of v **do**

    if Pred(w) = 0 (ie, w unreached)

    then Set Dist(w) to Dist(v) + Length(v,w)
      Set Heur_Dist(w) to Dist(w) + Euclid_Dist(w,y)
      Set Pred(w) to v
      if w = y **then return**
      Insert (Reached, w)

else if Member(Reached,w) **and*** Dist(w) > Dist(v)+ Length(v,w)

    then Set Dist(w) to Dist(v) + Length(v,w)
        Set Heur_Dist(w) to Dist(w) + Euclid_Dist(w,y)
        Set Pred(w) to v
        Change (Reached,w)

End_Procedure_Sedgewick_Vitter

3-5 FIBONACCI HEAPS AND DIJKSTRA'S ALGORITHM

We showed how to enhance the performance of Dijkstra's algorithm using heaps in Section 3-1. We can improve its performance even further using a sophisticated implementation of a heap called a Fibonacci heap. This leads to an $O(|V| \log |V| + |E|)$ implementation of Dijkstra's algorithm for a digraph $G(V, E)$. This improvement arises from the ability of a Fibonacci heap to realize the basic heap operations in $O(\log |V|)$ (average) time each, and the special change-key (strictly, decrease-key) operation required by Dijkstra's algorithm in $O(1)$ (average) time. The performance estimates of each of the Fibonacci heap operations will be given not in terms of worst-case times, but in terms of the so-called amortized time of each operation, which is essentially a running average of the time for the operation. Despite this, the performance estimate for the resulting implementation of Dijkstra's algorithm is still in terms of worst-case times. Thus, Dijkstra's algorithm uses the heap operations which require $O(\log |V|)$ amortized time at most $O(|V|)$ times each, and it uses the $O(1)$ change-key operation at most $O(|E|)$ times, whence the $O(|V| \log |V| + |E|)$ performance estimate follows.

    A heap is usually represented as a one-dimensional array. The representation of a Fibonacci heap is considerably more complex and is defined in terms of a collection of heap ordered trees. A *heap ordered tree* is a rooted tree with keys arranged in heap order: the smallest key is at the root of the tree, and the root of any subtree is the smallest key in that subtree. A *Fibonacci heap* is a disjoint collection of heap ordered trees. Each node u in a Fibonacci heap has four pointers: a pointer to the tree parent of u (which is nil if u is a root); a pointer to one of the children of u; and two other pointers that serve to embed u in a doubly linked circular list consisting of all the siblings of u. The roots of the separate heap ordered trees are also arranged in a doubly linked circular list. Each node u contains a field Rank(u) which gives the number of children of u. Although there is no a priori bound on the rank of any node, the heap operations are implemented in such a way that the maximum rank is at most $O(\log n)$ for a heap with n nodes. Each node also contains a mark bit used to facilitate certain of the heap operations. A separate minimum pointer points to the root containing the smallest key. Finally, just as for the heap representation described in Section 3-1, we assume we can directly access the nodes of the heap by an auxiliary array of pointers.

    We will consider the following heap operations: create, findmin, meld, insert, deletemin, decrease-key, and delete. The meld(h_1, h_2) operation combines the Fibonacci heaps h_1 and h_2 into a single Fibonacci heap. The Decrease-key operation decreases the key of an item in the heap by a positive amount. The Delete(h, i) operation deletes item i from the heap. Several of the operations are trivial to implement. Create merely returns a pointer to an empty heap in $O(1)$ time. Findmin merely accesses the minimum pointer in $O(1)$ time. Meld(h_1, h_2) combines the root lists of h_1 and h_2 into a single list and sets the new minimum to the smaller of the original minima in $O(1)$ time. Insert($\mathbf{h}, \mathbf{i}$) inserts element $\mathbf{i}$ in heap $\mathbf{h}$ by first creating a new heap $\mathbf{h}'$ consisting solely of item $\mathbf{i}$, and then melding $\mathbf{h}$ and $\mathbf{h}'$. The deletemin, decrease-key, and delete($\mathbf{h}, \mathbf{i}$) operations are more complex to implement.

Deletemin($\mathbf{h}$) deletes the minimum element from the Fibonacci heap $\mathbf{h}$. It first removes the node u containing the minimum (root) element. This leaves the children of u without a parent, and so the list of children of u is combined with the existing root list, making each child of u a new root. We then apply a series of linking operations on the resulting heap ordered trees, where each root determines a tree. Thus, if any two trees (roots) have equal rank, we link them by making the root with the larger key point to the root with the smaller key, breaking ties arbitrarily, and updating any effected fields, including the rank of the new root. This linking process is repeated until no two trees of equal rank remain. The new root list and the new minimum root can be identified on the fly. The worst-case time of this operation may be greater than $O(\log n)$. However, the amortized time of the operation, which we shall define shortly, can be shown to be $O(\log n)$. We can efficiently find and link trees of equal rank by augmenting the heap with an additional array of pointers $\mathbf{A}$, of size at most $O(\log n)$. The components of $\mathbf{A}$ are initially nil. As we scan the root list, and encounter a root u of rank i, we look up $\mathbf{A(i)}$. If $\mathbf{A(i)}$ is nil, we make $\mathbf{A(i)}$ point to u. If $\mathbf{A(i)}$ already points to the root v of another tree of rank i, we link the trees rooted at u and v, making the larger root point to the smaller root, creating a tree of rank $i + 1$, and setting $\mathbf{A}(i)$ to nil. We then continue the scanning and linking process by trying to insert the new tree at $\mathbf{A(i + 1)}$, iterating in this manner until we eventually create a tree T of some rank j not equal to the rank of any existing tree. At that point, we make $\mathbf{A(j)}$ point to the root of T. This process is repeated until all the rooted trees have roots of distinct rank.

We introduce the amortized time of a heap operation as a way of modeling the uneven performance of an operation whose average performance is much better than its worst case performance. First, we define the *potential* of a Fibonacci heap to be the number of trees it contains. The *amortized time* of a heap operation is then defined as the actual execution time of the operation plus the change in potential the operation causes. The *total amortized time* of a sequence of heap operations is the total actual execution time of the operations plus the net change they cause in potential. If the Fibonacci heap initially has zero trees, the net change in potential is necessarily nonnegative; so the total amortized time is an upper bound on the total actual time of the sequence of operations.

The amortized time of the create, findmin, meld, and insert operations are each $O(1)$, since the actual time of each operation is $O(1)$, and only the insert operation affects the potential, increasing it by 1.

The deletemin operation takes amortized time $O(\log n)$. For, it takes $O(1)$ actual time to remove the minimum element and combine its children with the other roots. Assuming the maximum rank is $O(\log n)$, this increases the potential by at most $O(\log n)$, so the amortized time may be $O(\log n)$. The actual time to perform k linking operations is $O(k)$, but each linking operation reduces the potential of the heap by 1; so k links reduce the potential by k, counterbalancing the cost in actual time. Consequently, the amortized time remains at most $O(\log n)$. Observe that the actual time of a deletemin may even be linear in n; for example, consider the very first deletemin operation after a sequence of insertions into an empty heap. But, the running average tracked by the amortized time is guaranteed to be at most $O(\log n)$.

It remains to describe the implementation of the decrease-key and delete($\mathbf{h}, \mathbf{i}$) operations, and to establish the logarithmic upper bound on the maximum rank of a tree.

To decrease the key of an item u by a positive amount, we first locate the node for the item, which we can directly find using the auxiliary array; decrease its key; break the pointer to its parent, if any; and adjust its parent's rank and child pointer. The node u becomes a root and may even be the new minimum root. All this takes $O(1)$ actual time. Later, we will introduce a refinement, which is necessary to ensure the $O(\log n)$ rank bound assumed in the analysis of deletemin. This refinement may trigger a chain of additional actions when the link between u and its parent is broken, but there will be a countervailing effect on a correspondingly modified version of the potential, so that the amortized time bound will still be $O(1)$. Delete($\mathbf{h}, \mathbf{i}$) is implemented similarly. We merely break the link between item $\mathbf{i}$ and its parent; combine the list of the children of item i with the original root list. This all takes $O(1)$ time, except if item i happens to be the minimum root, in which case the delete operation degenerates into a deletemin operation with $O(\log n)$ amortized time.

The refinement of decrease-key and delete needed to ensure the proper time bounds is as follows. A linking operation makes a node u a child of a node v. As soon as u loses two of its children, we apply the following rule: cut the link between u and its parent v, thus making u a root. Whenever this condition occurs, the induced cutting action can obviously trigger a chain of such cuts, called a *cascade of cuts*. A simple artifice allows us to implement the test for this situation efficiently. Thus, whenever we link a pair of nodes u and v, making u a child of v, we turn off the mark bit in the child u. On the other hand, whenever we cut the pointer from a node u to its parent v, if the parent v is an unmarked nonroot node, we turn on the mark bit in v (to indicate v has just lost one child); while if the parent v is a marked nonroot node (which indicates v has previously lost one child), we apply the cutting rule by cutting the edge between v and the parent of v.

The effect of the cutting rule is to ensure that the size of any tree in a Fibonacci heap is at least exponential in its rank, as demonstrated by the following theorem; f denotes the golden ratio $(1 + \sqrt{5})/2$.

Theorem (Fibonacci Rank Bound). A node of rank k in a Fibonacci heap has at least f^k descendants.

The proof is as follows. Consider the children of a node x of rank k in the order in which they became children of x. Observe that the ith child y of x has rank at least $i - 2$. For, by definition, when y was first linked to x, both x and y had the same rank, and, since x had rank at least $i - 1$ at that point, the rank of y must have been at least $i - 1$ at that point also. Subsequently, by the cascading cut convention, y could have lost at most one child without being cut from x. Therefore, since y is still a child of x, its rank must be at least $i - 2$. If we denote the minimum size of a subtree rooted at a node of rank k by S_k; then by the previous observation $S_k \geq S_0 + \ldots + S_{k-2} + 2$. The Fibonacci numbers F_k satisfy the recurrence relation $F_{k+2} = F_2 + \ldots + F_k + 2$, and so $S_k \geq F_{k+2}$. It is well-known that the Fibonacci numbers satisfy the inequality $F_{k+2} \geq f^k$, whence the theorem follows.

To complete the analysis of the heap operations, we need to modify the definition of the *potential* of a Fibonacci heap which we now define to be the number of trees in the heap plus twice the number of marked nonroot nodes. With this definition of potential, the amortized time analysis for the create, findmin, meld, insert, and deletemin operations is the same as before. For the decrease-key operation, we argue as follows: k cascade-cuts increase the number of trees by k but decrease the number of nonmarked nonroot nodes by k; so the net change in potential is $-k$. This counterbalances the actual time cost of the k cascade-cuts. Thus, the $O(1)$ amortized bound still holds. A more detailed description of the effect on the potential is as follows. Each operation increases the potential by at most 3 minus twice the number of cascade-cuts. This arises from the following contributions: plus 1 (for the new tree introduced by the first cut), plus 2 (since the root of the new tree may have been a marked nonroot node), plus 1 for each cascade-cut (each of which introduces a new tree), minus 2 for each cascade-cut (since each cut reduces the number of nonmarked nonroot nodes), for an overall change in potential of at most 3 minus the number of cascade-cuts. Thus, the amortized time is still $O(1)$, just as for the simpler situation before we introduced cascade cuts. The delete operation similarly takes $O(\log n)$ amortized time. The performance of the Fibonacci heaps is summarized in the following theorem.

Theorem (Fibonacci Heap Performance).   Let F be a Fibonacci heap which is initially empty, and suppose we perform an arbitrary sequence of heap operations on F. Then, the total actual time of the operations is at most the total amortized time of the operations. The amortized time for the deletemin and delete operations is $O(\log n)$. The amortized times for the remaining heap operations are $O(1)$.

Figure 3-18 illustrates the development of a Fibonacci heap. We assume the heap is initially empty and that elements with keys equal to 1 through 9 are inserted.

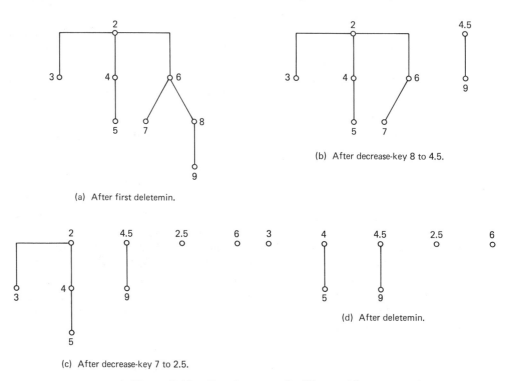

(a) After first deletemin.

(b) After decrease-key 8 to 4.5.

(c) After decrease-key 7 to 2.5.

(d) After deletemin.

Figure 3-18.   Development of a Fibonacci heap.